MARITIME WEATHER AND CLIMATE

FIRST EDITION

WILLIAM J. BURROUGHS
and
NORMAN LYNAGH

British Library Cataloguing in Publication data
Burroughs, William J. and Lynagh, Norman
Maritime Weather and Climate
1. Marine
1. Title

ISBN 1 85609 166 X

MARITIME WEATHER AND CLIMATE

FIRST EDITION

WILLIAM J. BURROUGHS
and
NORMAN LYNAGH

LONDON
WITHERBY & CO LTD
32-36 Aylesbury Street
London EC1R 0ET

First Published 1999

WITHERBY

PUBLISHERS

© William J. Burroughs
Norman Lynagh
1999

ISBN 1 85609 166 X

Published and Printed by:
Witherby & Co. Ltd
32-36 Aylesbury Street
London EC1R 0ET

Tel: 020 7251 5341
Fax: 020 7251 1296
International Tel: +44 (0)20 7251 5341
International Fax: +44 (0)20 7251 1296
E-mail: books@witherbys.co.uk
www: witherbys.com

PREFACE

It is surprising how little has been published for mariners and the maritime industry about recent developments in weather and climate matters. There are plenty of books on the basic meteorology and how this affects the weather at sea. But up-to-date analysis on the wider features about the conditions, which have to be confronted on the high seas, how new technologies can monitor and predict these conditions, and whether the climate is shifting so as to pose new challenges for maritime operations, is not to be found in a single accessible form. Indeed, some of the more complete texts of conditions around the world, such as the Admiralty Weather Manual, date from before World War II.

Against the background of the growing public awareness of the global nature of the climatic events, as highlighted in the all-pervading influence of the El Niño, and the mounting concern about more extreme weather imposing unmanageable burdens on the insurance industry, this limited, and often dated information makes it difficult to make informed decisions about the challenges facing many parts of the maritime industry. This book is designed to address this gap.

But how substantial is this gap? The answer lies in the pace of development of emerging technologies. Satellites and automatic buoys, which are providing new insights in to everything from instantaneous events including extreme waves, storm surges and explosive storm development to providing improved climatologies of conditions around the world. Supercomputers are transforming forecasting services and producing models which may provide guidance on the changes in the climate at sea over the next century. Clearly, we cannot cover in detail the technicalities of all these immensely complicated endeavours.

v

Instead, our objective is to provide an accurate, balanced presentation of where these various activities have got to and to provide guidance on how to approach the avalanche of services that is being marketed to maritime industries.

In seeking to give an up-to-date and forward-looking assessment of current developments, it is important to maintain a sense of historical perspective, especially in terms of injecting a sense of realism into any assessment of changes in the incidence of extreme events. This is a matter of achieving a judicious balance. On the one hand we cannot bury our heads in the sand and insist that there is nothing new under the Sun, and that we have seen it all before. On the other hand, we must be careful to avoid the trap of being swept away by the rhetoric of those who, in trying to sell a service, may wish to present current events in an alarmist fashion. So there is a real value in presenting plenty of examples of past extreme events to get a better feel for whether things are really changing, and, if so, by how much.

The overall aim of this general approach is to enable the reader to become better versed in the current status of knowledge of maritime weather and climate, and so make better use of the opportunities services provide to reduce the threats posed by changes in the weather and climate. A measure of the success of this book will be how many readers seek to find out more about the science and technology that underpins these services and so discover their strengths and weaknesses in protecting maritime industries from the perils at sea.

William Burroughs
Norman Lynagh

August 1999

ACKNOWLEDGEMENTS

We would like to thank the following people for help in providing advice, helpful discussion, background papers and copies of figures: Tim Corbett, John Doody, Alan Douglas, Chris Folland, Jack Hopkins, Stuart Norwell and David Parker at the UK Meteorological Office, David Anderson, Tony Hollingworth, Peter Jannsen, Tim Palmer and Austin Woods at the European Center for Medium Range Weather Forecasting, Grant Bigg, Tom Holt, Mike Hulme and Phil Jones at the University of East Anglia, David Cotton at the University of Southampton, John Thompson at Oceanroutes (UK) Ltd, Christopher Landsea at NOAA Hurricane Research Division, Miami, David Blackman at the Proudman Oceanographic Laboratory, Chris Mutlow at the Rutherford Appleton Laboratory and Neil Lonie at the University of Dundee.

ABOUT THE AUTHORS

DR WILLIAM JAMES BURROUGHS spent seven years researching atmospheric physics at the National Physical Laboratory, Teddington, England, before working as a Scientific Attaché at the British Embassy in Washington. Between 1974 and 1995 he worked in a variety of senior policy posts in Whitehall for the UK Department of Energy and then Health. He is now a successful popular science writer and has had a variety of books published on the weather including *Does the Weather Really Matter?*, *Weather: the Ultimate Guide to the Elements*, *Mountain Weather: a Guide to Skiers and Hillwalkers*, *Weather Cycles: Real or Imaginary?* and *Watching the World's Weather*.

NORMAN LYNAGH is a Chartered Meteorologist who bagan his career in the UK Meteorological Office in the early 1960s. His six years there included three years' service on the North Atlantic Ocean Weather Ships.

He took up a post with the Australian Bureau of Meteorology in 1967 and trained as a Forecaster. Following that training, he spent three years working in Queensland, the Northern Territory and New Guinea, forecasting primarily for aviation and the general public.

In 1970 he joined Imcos Marine Ltd, a Britih company that specialised in the provision of marine weather forecast services to the offshore oil industry all over the world. In his 10 years with the company he worked in all continents except South America, often working at sea on board drillships, semi-submersible drilling rigs, crane barges and pipelaying barges.

His next career move was to the Noble Denton Group, one of the leading international marine consultancy firms. Initially the work was primarily related to the calculation of meteorological data for use in the design of offshore structures but soon he led the establishment of a marine weather forecasting service. In the mid 1980s Noble Denton Weather Services was formed as a separate operating company within the Group. Norman Lynagh was appointed Managing Director in 1986 and remained in that position until he left the Noble Denton Group in 1996. In his later years with Noble Denton he specialised in Forensic Meteorology, working on detailed examination of past weather events to assist in litigation and arbitration.

In 1996 he became a founding Partner in Norman Lynagh Weather Consultancy, which provides forensic meteorology services to the legal and insurance professions and weather forecasting services to the marine industries.

In 1989 Norman Lynagh was the Meteorologist on Tom Gentry's team when it beat the trans-Atlantic speed record with the *Gentry Eagle*. In 1998 he was the Meteorologist on the team that beat the round-the-world record with the *Cable & Wireless Adventurer*.

CONTENTS

List of Figures

CHAPTER 1

INTRODUCTION

'As when to them who sail
Beyond the Cape of Hope and now are past
Mozambic, off at sea north-east winds blow
Sabean odours from spicy shore
Of Araby the blest, with such delay
Well pleased they slack their course and many a league
Cheer'd with the grateful smell old ocean smiles'

John Milton,
Paradise Lost

Ever since men started crossing the oceans in boats, the weather has played a vital role in sea-faring. So a knowledge built up among sailors of the ever-changing nature of the weather and the underlying features of the climate with its prevailing winds and ocean currents. The reference to Sabean odours in Paradise Lost, like the legends of Sinbad the Sailor, reflects how this knowledge developed in the Indian Ocean and spread around the world. The Sabean odours refer to the production of frankincense (the aromatic resin of the gum trees of the genus Boswellia) in the coastal strip of southern Arabia.[1] The gum trees in this region, which was ruled by the Sabeans of Ophir, are strongly scented and in winter the dry northerly wind carries these odours out to sea. This consequence of the winter monsoon wind patterns was known to ancient sailors and picked up by European mariners when they first ventured into the Indian Ocean in the 16th century.

This knowledge was evident in the huge range of hazardous trips taken by ancient sailors in maintaining trade routes and colonising far-flung parts of the world. In extreme cases the climate could change sufficiently to disrupt these links. For example the collapse of the Norse colony in Greenland was hastened by more extensive sea ice in the 13th and 14th centuries which prevented trade with Iceland continuing. But, almost everywhere sailors developed the skills needed to handle the weather, albeit at considerable risk to themselves, their cargoes and their craft. By the mid-19th century, the most successful shipmasters of the 'clipper' age were weather experts. Their ability to use meteorological and ocean current knowledge to obtain the fastest passage halfway round the world, with the minimum risk and high profitability is a measure of their great skill.

The coming of steam ships changed the situation. Because progress depended so much more on the performance of the vessel the importance of weather knowledge declined and a more resigned approach entered into mariners' thinking. Indeed there was a sense that technical progress made the weather less important, and it was with tragedies like the sinking of the 'Titanic' in April 1912 that this assumption was brought into question. At the same time the introduction of radio communications transformed the attitude to the weather. Although the 'Titanic' was equipped with radio and received reports from other ships about the extensive ice conditions south of Newfoundland during its fateful trip, it was the introduction of radio bulletins for shipping after the First World War that really started to redress the balance.[2]

1.1. Warnings and Forecasts

The scale of the 'Titanic' disaster galvanised international action to establish warning systems for ice conditions in the western North Atlantic. These have been built round ice-patrols maintained by the US Coast Guard since 1914, except during the Second World War. Alongside the development of improved monitoring arrangements and advances in radio communications providing bulletins about conditions at sea , the science of weather forecasting was making steady progress. Indeed the establishment of weather forecasting organisations around the world was closely linked to disasters at sea from the mid-19th century onwards. The devastating storm that hit the Anglo-French fleet in November, 1854, in the Crimea destroying 13 transporters laden with food and clothing for the troops and forage for the horses caused far more hardship, disease and loss of life during the subsequent winter than did military action. These losses were instrumental in setting up both the British and French weather services. Similarly, the storms that hit the Great Lakes in the

United States in 1868 and 1869, sinking or damaging over 3000 vessels and killing 500 people led to the establishment of the US Weather Bureau.

The early work of these services sought to exploit newly-available measurement and communications systems to provide warnings of damaging storms. This approach relied on telegraphy, which enabled information about current weather conditions to be collected and used to provide some warning of advancing storms. In the UK, following the sinking of the steam clipper *Royal Charter* off the coast of Anglesey in October 1859 with the loss of 400 lives, the Meteorological Department of the Board of Trade started issuing storm warnings in February 1861. These were transmitted by telegraphy to coastal stations, where a system of 'cautionary signals' using cones and drums by day, and lights by night, was hoisted from a mast and yardarms, for shipping to see, when gales and storms were expected. Unfortunately, after a few years, the poor performance of these forecasts drew adverse criticism and they were abandoned between 1867 and 1874.

Eventually improved observation networks enabled meteorologists to build up a better picture of weather systems. This led to advances in the science in meteorology during the early decades of this century, and increasingly useful forecasts were produced. But, the value to shipping was limited, as the amount of information that could be conveyed in radio broadcasts was limited, given the huge areas any forecast covered. While radio-coded analyses from meteorological centres became available in 1948, which enabled ships' officers to construct weather maps, this was a slow progress which required considerable skill. So the subsequent introduction of radio-facsimile chart transmission represented a major step forward. A large proportion of the world's merchant fleet now use this technology as a primary source of weather information. It enables ships' masters to get a clearer picture of how the weather is evolving than can be obtained from text weather forecasts.

The part played by technology is even more crucial in the progress of weather forecasting. Ranging from the world of supercomputers to weather satellites these developments have provided meteorologists with the tools to make considerable advances. This has led to the provision of increasingly sophisticated services to maritime industries, which will be discussed at length later in this book. In part these services reflect the growth of these industries and their needs. As the range of vessels and offshore installations has widened their successful operation has depended on increasingly accurate predictions of adverse weather and, in particular, identifying periods of favourable weather - *weather windows*. This now involves improved routeing for the most economical voyages, through choosing the safest time to float-out major installations, to taking evasive or precautionary action when severe storms threaten.

All these services are now an essential feature of marine operations and an improved understanding of what is available is part of good business practice. Indeed, many charterers stipulate that shipping operators must use approved routeing services if they are to get contracts. This requirement is not simply a matter of improving voyage times but addresses the vexed question of whether ships match up to their claimed performance. In the past it was all too easy for the master or owner to claim that weather conditions resulted in delays when, in practice, the ship was not up to fulfilling its contract. Now the standard routeing analysis enables charterers to check whether the weather was to blame or whether any delays were a consequence of the ship not being capable of meeting its contractual conditions.

In exploiting these services it pays to have some idea of how the forecasts are prepared and their strengths and weaknesses. In spite of frequent criticism, weather forecasts have made substantial progress in recent decades. These advances are the product of developments in various areas. First, and foremost, the immense increase in the power of computers which has enabled the meteorologists to develop more detailed models of the atmosphere and carry out the vast number of calculations required to prepare better forecasts. Secondly, there has been a steady advance in our physical understanding of how the dynamics of the atmosphere and oceans work, and then how the models can be improved to reflect this knowledge. Thirdly, there have been many improvements in how the properties of the atmosphere and oceans are measured so that better data is fed into the forecasting models. These developments have included not only better instrumentation, the increasing use of automatic equipment on aircraft, buoys and ships, and the whole area of weather satellites. Nevertheless, the number of observations at sea have declined as there are now fewer ships plying the oceans, and this, combined with a shortage of surface observations in some developing countries and sparsely populated areas, means there are still substantial gaps in the observing network.

How progress has been made on all these fronts provides not only insight into the challenges facing forecasters and anyone who uses their predictions, but also why there will also be limitations in what can be done. So the workings of weather forecasting will be explored in considerable detail with the aim of helping users at sea to appreciate both the value of forecasts and their inherent limitations. This will cover not only the exploitation of forecasts to increase the efficiency and effectiveness of operations but also to address the issues of negligence raised by failure to take adequate safety precautions to avoid the consequences of predicted adverse weather. This is a contentious area. There are a sizeable number of disputed insurance claims, some of them involving sums of money well in excess of £10 million ($16 million), which do end up in litigation. In most cases a form of settlement is reached rather than a judgement against one party or another. In these circumstances, an accurate assessment of the part played by adverse weather is central to reaching a judgement.

Another area of contention is the balance of responsibility between charterer and the owner when ships are required to enter ports, which do not have adequate forecasting services, and a predictable weather event causes damage, For example, at a recent arbitration an exposed port was deemed by the tribunal to be unsafe on the grounds that it did not have a local site-specific weather forecasting service. For that reason, the charterers were found to have been in breach of the charter agreement by ordering the vessel to unload at an unsafe port.

1.2. Climatology

The planning and operating of maritime businesses to take account of the fluctuations in the weather and the climate is not just a matter of being prepared for adverse events. It also requires a thorough background knowledge of normal conditions, and how these can fluctuate about the normal. While much of what will be considered in this book will be extreme events (e.g. hurricanes, intense extratropical depressions, extreme waves etc), the fact that these are part and parcel of the range of normal conditions is central to a proper understanding of hazardous weather at sea. So, it is essential to have an adequate knowledge of the statistics of what can be expected around the world.

The climatology of the oceans has been built up from the painstaking observations of mariners over the centuries. Because most of these data have been collected on the principal trade routes they provided only a spotty picture of conditions around the world (Fig 1.1). Although many of these gaps were, over time, partially filled in by measurements from other voyages, involving trade, exploration and scientific research, the climatologies built up from these observations were far from complete. An early example of these limitations was the difficulties with heavy swell experienced during the military landings on the Atlantic coast of North Africa in November 1942. It became clear that very little was known about the physical properties of ocean waves and their propagation over long distances, and a research team was set up at the Admiralty Research

Laboratory in England to develop measurement techniques to study the distribution of wave periods and heights. This work acted as a stimulus to developing the oceanography after the war. It soon became clear that many areas needed more study. Making accurate measurements of wave characteristics not only provided insight into their behaviour but also showed that ocean swells could travel extraordinary distances with little loss of amplitude.[3] This remains an important maritime hazard. A central feature of a recent arbitration case was the sudden arrival of enhanced swell energy at a Moroccan port.

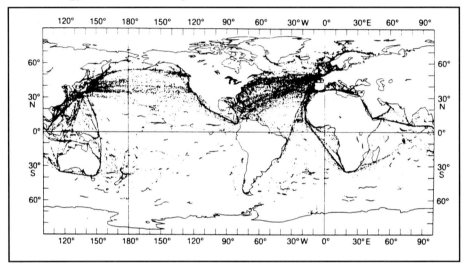

Figure 1.1
Distribution of routine in situ (ship and buoy) sea surface temperature data for October 1986. (From Gurney et al, 1993, with permission of Cambridge University Press.)

The expansion of physical oceanographic work after the Second World War has led to substantial improvements in our knowledge of the climatology of the oceans. This work was supplemented by major studies, such as the Joint North Sea Wave Project (JONSWAP),[4] which greatly helped to improve knowledge of the frequency spectrum of ocean waves. When combined with observations from shore stations, light vessels, and ocean weather ships, these measurements proved invaluable when it became necessary to design oil platforms in the North Sea and elsewhere.[5] Nevertheless, these specific measurements, together with the restricted coverage of normal shipping, mean that there are still major limitations with the standard climatologies. This is not surprising because, while we may have a reasonable picture of average conditions, the range of extremes, which can normally be expected in any given location, will never be adequately covered by occasional visits. Only by having regular, standard measurements of the most relevant parameters (e.g. atmospheric pressure, temperature, wind speed and direction, wave height and extent of sea ice) is it possible to provide a reliable measure of the range of conditions that can occur at any time of year, and whether this range has shifted significantly over the years.

Standard shipboard observations are not the answer to the shortcomings in published climatologies of the oceans. They will not be able to provide adequate coverage. Moreover, in some instances the measurements are simply not good enough. Both wind speed and wave height observations have significant limitations. So, while these measurements continue to play an important part in developing maritime meteorology and climatology, real improvements in building up a global picture of conditions at sea depends on new technologies. Two areas are making major contributions. Automatic moored and drifting buoys are monitoring the most relevant areas of the oceans with more detail and precision than is possible on most ships. The new generation of oceanographic satellites (e.g. Seasat, Geosat, Topex/Poseidon & ERS 1 & 2) have transformed our ability to measure meteorological parameters over the oceans.[6] In addition, the improved analysis provided by the computer analysis of measurements used as the input to the numerical forecasting of the weather has produced a better picture of conditions in remote parts of the world.

New technologies do more than fill in the gaps in the climatologies. In some instances they are showing that existing data bases may be misleading. Wave height measurements are showing that the single maximum waves in severe storms may be much greater than the published statistics indicate. This has substantial implications for the design and operation of ships, off-shore platforms and coastal structures. So here we will seek to combine the essence of the standard climatologies that have been built up by mariners over the years with the rapidly emerging results of the measurements made by buoys and satellites which are providing many new insights into both the behaviour of the surface of the oceans and the weather at sea. The explosive growth in the data being produced means there is an urgent need to synthesise the core information into a form which is useful to mariners. Here we will attempt to distil the essence of this rapidly developing area.

1.3. Climate Change

Against the background of establishing an adequate climatology of the oceans, there is the intriguing question of how the climate is changing. At a time of growing concern about global warming and its implications for extreme events at sea, there is good reason for knowing about such changes. In particular, the rising cost of insurance losses resulting from all forms of weather events (e.g. drought, floods and storms) has led to widespread claims of a rising incidence of extreme events. These figures do need, however, to be treated with great care. Any analysis of insurance losses must take full account of changing levels of insurance cover, population densities in vulnerable areas (e.g. coastal sites), inflation, and even fraudulent claims. Equally, claims of a rising incidence of extremes is often linked to the belief that, as a consequence of global warming, hurricanes[7] are becoming more intense and more frequent. In practice the figures for the tropical North Atlantic show, if anything, the reverse is true (Fig. 1.2).[8]

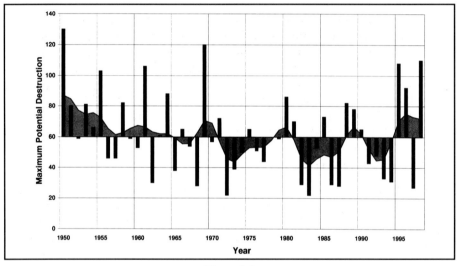

Figure 1.2
The maximum potential destruction by hurricanes in the North Atlantic between 1950 and 1998. This measure provides a way of estimating the destructive power of the hurricanes in any year by combining the number of days when wind speeds were above a given level with the square of the wind speed to reflect the fact the most intense hurricanes do by far the most damage. This index shows a marked decline until the early 1990s, as do other measures of hurricanes in the Atlantic. (Data from the University of Colorado.)

Analysis of shipping casualties requires the same caution. Figures prepared by Oceanroutes Inc. for weather-related losses during the period 1967 to 1975, drawing on data published by Lloyds Register, concluded that 138 ships of 5000 tons dead-weight or greater had been lost of which 83 per cent were the result of heavy weather or tropical storms, and 16 per cent were due to fog or poor visibility (the remaining ship was struck by lightning). This analysis, together with more recent work, published in 1985,[9] concluded that after rising sharply during the

1960s and 1970s, the overall number of ships lost had stabilised in the 1980s. Weather-related losses had risen as a proportion of total, until 1980, and then settled down at between 20 and 25 per cent. So for the purposes of this book our starting point will be that heavy weather and poor visibility are constant hazards for maritime operations. Once we have explored what is known about various extremes we can then consider the implications of their frequency changing.

The consequences of climatic change do not relate only to more frequent extreme events but are more wide-ranging. Some views of a future warmer world predict, say, a decline in the winter westerlies in mid-latitudes. If this also leads to less intense depressions and lower peak wind speeds, there could be significant implications for both the operation and design of both vessels and off-shore installations as lighter, more flexible systems could function safely in such conditions. So it pays to be up-to-date on thinking about climatic change, both in terms of what broadly is happening and what this could mean for the future incidence of extreme events. In this context, the recently completed monumental work of the Intergovernmental Panel of Climate Change (IPCC)[10] provides an excellent opportunity to take stock of current knowledge of how the climate is changing and what this means for the future.

1.4. Shipboard Observations

All this emphasis on scientific developments, new technologies and global perspectives should not disguise the basic importance of the practical reality of dealing with daily weather and, in particular, the value of shipboard observations. Two essential features of these measurements will be explored in this book. First, the making of regular and reliable measurements of the weather increases awareness of current meteorological developments and hence improves the capacity to interpret forecasts and related information effectively. Secondly, where part of a formal programme, accurate observations can make a valuable contribution to building a better data base for both day-to-day forecasting and longer term climatological analysis, which is of mutual benefit to both maritime organisations and weather services. So this aspect of maritime weather and climate will be given prominence, without rehearsing the details of making standard measurements which is covered in the manuals.[11]

1.5. A Maritime Perspective

To put the information provided for mariners by various forecasting and advisory services to practical use requires a grounding in basic meteorology. Much of this is available in standard textbooks. But, there is an essential maritime flavour, which is not always easy to extract from these texts, and this needs to be addressed before embarking on more detailed exploration of the subject. Moreover, while there are a few standard texts on the more basic aspects of maritime weather and climate,[11, 12] the wider implications of current weather and climate developments are not covered from a marine point of view.

In these days of global computer models for both numerical weather prediction and studying climate change, it could be argued that separating out the maritime aspects of meteorology is an artificial distinction. Why do we need another book on meteorology and climate change, when so many of the aspects of the subject are covered in the existing texts on the subject? The answer is that virtually all of these books have presented the issues from a land-based perspective. This is hardly surprising as most meteorologists live and work ashore as do the vast majority of the users of their services. So, while many books recognise the fundamental role of the oceans in weather and climate, the application of this knowledge is applied principally to land-based examples.

At the most basic level the essential features of weather systems (e.g. mid-latitude depressions, anticyclones, and hurricanes are adequately described in standard texts (see Bibliography). Furthermore, the fundamental part played by the oceans in driving and modifying these systems is well covered. Where the land-based bias becomes apparent is in more detailed aspects of behaviour, notably in three areas.

First, in the westerly flow in mid-latitudes of the northern hemisphere, many books are inclined to lose sight of weather systems after they have crossed the land masses. This is particularly true of those which concentrate on North America. But, as depressions move from the continents out into the Atlantic or the Pacific, the new vigour they can sometimes gain in complex interactions with the warm water of the western boundary currents (i.e. the Gulf Stream and the Kuro Shio) is of real interest to mariners. Similarly, the properties of arctic air masses over eastern Asia and eastern North America change dramatically as they cross these warm waters. The marine perspective, and its implications for safety at sea, is rarely covered in the standard texts.

Another aspect of this westerly flow, which looks rather different to those on-board ship as opposed to onshore, is the broad circulation patterns steering depressions around the globe. As we will see in Chapter 2 these patterns are linked closely to strong winds in the upper atmosphere (the *jet stream*). The land-based perspective of these patterns is dominated by how they can produce either spells of mobile changeable weather or periods when conditions remain unchanged for days and even weeks on end. To the mariner voyaging around the globe, with or against the flow of these patterns, geographical distribution of their waveform is crucial to ocean routeing. So the marine perspective requires a clearer understanding of how the movement of weather systems are part of global patterns, and what this means for planning lengthy voyages.

A second, more widely recognised feature of ocean-atmosphere interactions is how the intensity and movement of storms, both tropical and extra-tropical, are fuelled and influenced by patterns of sea surface temperatures. These processes involve not only the major storm tracks, but also more localised effects such as the way developing depressions can run up the English Channel or down the North Sea, and the areas of cyclogenesis such as the Gulf of Genoa and the Skaggerak. Some aspects of these interactions are covered in other texts. They tend to focus, however, on issues which relate to when and where these systems will make landfall. Here what matters is the developing situation at sea.

The third area is the subject of much wider interest, but is developing so rapidly that it has to be covered here. This is the whole question of global ocean-atmosphere weather teleconnections and how they drive longer term features in the climate. Much of this work has centred on what is known as the El Niño Southern Oscillation (ENSO) which governs ocean and atmosphere circulation across the tropical Pacific. These patterns play a central role in seasonal weather patterns across the tropics and exert a more shadowy influence at higher latitudes.[13] As such they are of great interest to mariners, notably in terms of the part they play in the incidence of hurricanes, and whether useful seasonal forecasts can be made of their numbers and intensity.

So in presenting a review of basic meteorology and climatology, the emphasis will be on teasing out those aspects which are of most relevance to the safe and efficient operation of shipping. To avoid repeating much of what can be found in standard textbooks (see Bibliography), the presentation will concentrate on how to apply meteorological knowledge to exploit the services available now to the shipping industry. In particular this will focus on the impact of new technologies, services, and the emerging understanding of how the climate functions to provide:

a) ships' masters and crew with the basic information to make improved decisions on how to optimise day-to-day sailing;
b) shipping operators with a better understanding of the challenges the weather poses for the continuing efficient and safe operation of their vessels; and

> c) the insurance industry with a wider appreciation of the risks the weather and climate change presents for shipping and offshore installations now and in the future.

FOOTNOTES

[1]   Hatch (1985).
[2]   Murray (1992).
[3]   Barber & Ursell (1948).
[4]   Hasselmann *et al.* (1973).
[5]   Charnock (1985).
[6]   Young & Holland (1996).
[7]   Throughout this book the term hurricane will often be used to cover all tropical storms that exceed a specified level (sustained wind speeds of 64 knots or greater) including cyclones and typhoons, in spite of the fact that in the Indian and Pacific Oceans they are usually known by these alternative names. More generally, the expression tropical cyclone will be used to cover all tropical storms that have an identifiable circulation pattern (see Sections 2.6 and 3.9).
[8]   Landsea (1993).
[9]   Quayle (1986).
[10] IPCC (1995).
[11] See for example UK Met Office, (1978), and UK Met Office, (1995).
[12] Cornish & Ives, (1997).
[13] Philander (1990).

CHAPTER 2

BASIC METEOROLOGY

The Sun came up upon the left,
Out of the sea came he!
And he shone bright, and on the right
Went down into the sea.

Samuel Taylor Coleridge
Ancient Mariner

The oceans play a fundamental role in many aspects of both the Earth's climate and the day-to-day weather. Understanding how the oceans participate in both the wider climatic processes and in establishing local conditions is central to getting to grips with maritime weather and climate. But, before we can get down to these practical issues, there are a number of basic features of the Earth's climate which need to be addressed. The objective of this review is to identify those features of meteorology which are of greatest importance to maritime industries, so as to enable the reader to interpret more easily both standard texts on the subject and what follows in later chapters.

2.1 The Earth's Energy Balance

At the most basic level the Earth's climate is governed by the balance between the energy the planet receives from the Sun *(solar radiation)* and the amount of heat it radiates *(terrestrial radiation)* to space. Since the amount of terrestrial radiation is defined by the temperature of the Earth's surface and the atmosphere above it, the average temperature of the planet stays at a level to maintain a balance between incoming solar radiation and outgoing terrestrial radiation.

At this stage, it is helpful to consider the amount of solar radiation falling on the Earth as a whole as remaining constant. Although there are small changes associated with the Earth's elliptical orbit about the Sun, and even smaller changes associated with cycles of solar activity, these do not affect the analysis here. But, while the amount of solar radiation falling on the Earth is effectively constant, the amount falling on different parts of the globe is controlled by both the rotation of the Earth and the cycle of the seasons and goes through predictable diurnal and annual cycles. Also the amount absorbed varies greatly from place to place. This is because some of the solar radiation is either absorbed or scattered by the atmosphere and the particles and clouds in it. The remainder is either absorbed or reflected at the Earth's surface: the amount depending on the surface properties. Snow reflects a high proportion, whereas the oceans absorb solar radiation efficiently.

The amount of solar radiation reflected or scattered back into space without any change to its wavelength is defined as the albedo of the surface. The mean global albedo is about 30 per cent. The albedo of different surfaces can vary from 90 per cent to less than 5 per cent (see Table 2.1). The reflective properties of the ocean surface are dependent on the angle at which the sunlight strikes the surface and the wave structure of the surface. But it is clouds which make the biggest contribution to the Earth's albedo, as they cover about half of the surface at any one time. For this reason, both the type of cloud (see Table 2.1) and the amount of cloud cover are crucial in controlling how much of solar radiation is reflected into space and hence in defining the Earth's temperature. So our knowledge of clouds limits our ability to model the Earth's climate and to predict how it might change when perturbed by human activities or natural fluctuations.

Of the total incoming flux of solar radiation about 30 per cent of the incoming radiation is reflected or scattered back into space and about half penetrates the atmosphere and is absorbed by the Earth's surface. The rest (some 16 per cent) is directly absorbed by the atmosphere (Fig. 2.1). How the solar radiation is absorbed at the Earth's surface differs profoundly on land and at sea. On land most of the energy is absorbed close to the surface, which warms up rapidly, and this increases the amount of terrestrial radiation leaving the surface. At sea the solar radiation penetrates much more deeply, with over 20 per cent reaching a depth of 10 metres (33 feet) or more. So the sea is heated to a much greater depth and the surface warms up much more slowly. This means more energy is stored in the top layer of the ocean and less is lost to space as terrestrial radiation. This absorptive capacity of the oceans plays an important part in the dynamics of the Earth's climate. When combined with the fact that the oceans cover 71 per cent of the Earth's surface, have a thermal capacity of more than a 1000 times that of the atmosphere, and can transport heat between its surface and lower layers as well as horizontally, it is clear how central this climatic role is.

TABLE 2.1
The Proportion of Sunlight Reflected by
Different Surfaces (Albedo)

TYPE OF SURFACE	ALBEDO
Tropical Forest	0.10-0.15
Woodland - deciduous	0.15-0.20
- coniferous	0.10-0.15
Farmland/natural grassland	0.16-0.26
Bare soil	0.05-0.40
Semi-desert/stony desert	0.20-0.30
Sandy desert	0.30-0.45
Tundra	0.18-0.25
Water (0-60°)*	less than 0.08
Water (60-90°)*	0.10-1.0
Fresh snow	0.80-0.95
Sea ice	0.30-0.40
Snow-covered vegetation	0.25-0.60
Snow-covered ice	0.20-0.80
Clouds - low	0.70-0.90
- middle	0.65-0.75
- high (cirrus)	0.45-0.50
- high (cirrus)	0.15-0.25
- cumuliform	0.65-0.75

* The closer the sun is to the zenith the less sunlight is absorbed. Also the presence of whitecaps on the surface increases the albedo.

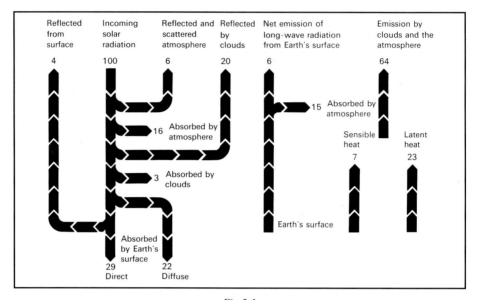

Fig 2.1
The total amount of solar energy falling on the Earth is balanced by the amount that is reflected or absorbed and re-emitted as terrestrial radiation (with permission of Cambridge University Press).

Most surfaces radiate heat efficiently, emitting at least 90 per cent of the energy of which they are theoretically capable. So for all practical purposes we can ignore any differences in the surface when considering how much energy is being radiated to space by any part of the globe. What really matters is the temperature.

The requirement to balance absorbed solar radiation and emitted terrestrial radiation for the Earth as a whole is the driving force behind the transport of energy by the atmosphere and the oceans. On a global scale the most important feature is that at high latitudes despite lower temperatures (typically -30°C in polar regions in winter), large amounts of energy are radiated to space. This means more energy is radiated to space than is received from the Sun (Fig. 2.2). This process is greatest in winter, when there is little or no solar radiation reaching these regions. This loss must be balanced by energy transport from lower latitudes. Even though this energy transport is reduced at other times of the year, it provides the motive forces to drive the global atmosphere and oceanic circulation throughout the year.[1]

The transport of sensible heat in the atmosphere and the oceans is only part of the story. Because the latent heat of vaporization of water is so high, the evaporation of water over the tropical oceans transports huge amounts of energy polewards when it condenses at higher latitudes. So explaining how the processes of evaporation and precipitation (the hydrological cycle) are bound up in the transport of energy is all part of understanding the workings of the global climate.

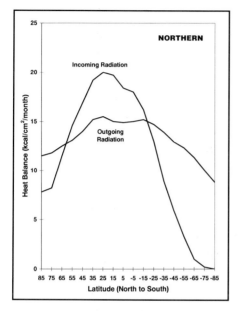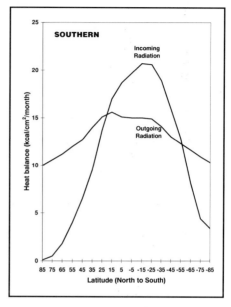

Fig 2.2a, and b
In general the amount of solar energy absorbed by the Earth at each latitude differs from the amount of terrestrial radiation emitted at the same latitude, and so energy has to be transferred from equatorial to polar regions.

2.2 General Circulation

How the atmosphere and the oceans transport energy around the globe is the key to understanding the climate. All the motions of both the atmosphere and the oceans are part of the fundamental process of maintaining the Earth's energy balance, and, in particular, transporting energy from equatorial to polar regions. In the case of the atmosphere, if the Earth was covered by only oceans, the long-term patterns of wind and temperature would show nothing but zonal bands with no longitudinal variation. Furthermore, if it did not rotate, the energy transfer would involve a simple meridional circulation with air rising at the equator, flowing to the poles, descending and returning at low level to the equator. With a rotating Earth this motion involves horizontal vortices and waves. The distribution of the oceans and the continents across the globe makes the motions yet more complex, but the underlying pattern retains many features of the simple zonal model.

The most basic representation of wind systems close to the Earth's surface is to regard them as a set of zones. The tropics feature easterly trade winds converging on the Equator, on either side of these there are calm subtropical high pressure zones. The mid-latitudes are dominated by westerly winds and the polar regions are areas of stormy low and high pressure zones. These can clearly be seen in Fig 2.3

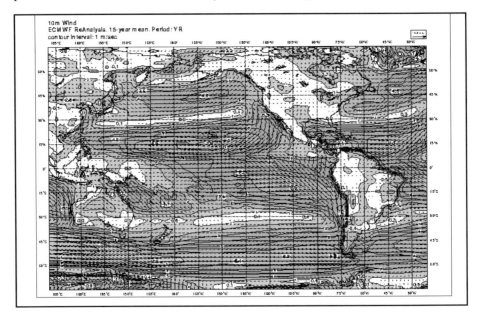

Fig 2.3
The principal wind zones around the globe can be seen in this climatological map (with permission of ECMWF.)

When the vertical motions are included the reasons for these zones become more obvious (Fig. 2.4). The weather in the tropics is dominated by vertical motion. This is driven by the large input of solar energy throughout the year. Known as the Hadley Cell, after George Hadley, the 18th century London lawyer who first proposed this atmospheric motion, the circulation consists of air rising near the equator to heights of up to 20 kilometres (65 000 feet) and then spreading out to 20 to 30°N and S latitudes before descending and flowing back toward the equator as the easterly trade winds in equatorial regions. These winds are of major climatic importance as they blow over nearly half the globe. Furthermore they are one of the most constant features of the Earth's weather system. Their average is around 14 knots (Force 4), but, are stronger in the winter as part of the greater energy transport in each hemisphere at this time of the year, and weaker in the summer.

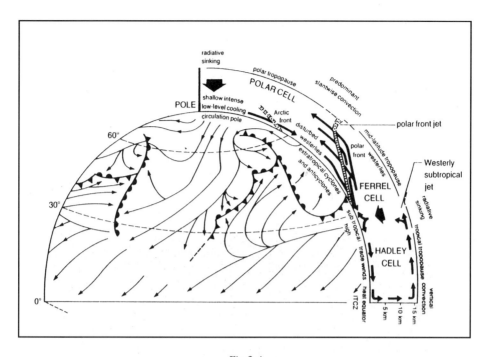

Fig 2.4
A schematic model of the general circulation of the northern hemisphere (in cross-section) showing the locations of the principal surface features, the vertical circulation, the polar front jet stream and the westerly subtropical jet. (From Musk, 1988, with permission of Cambridge University Press.)

Because much of equatorial regions are covered by warm oceans the rising air in the Hadley Cell is very humid. It cools as it rises. Initially the cooling rate is only about 3·5°C per kilometre as a lot of heat is released as the water vapour in the air condenses. As the air becomes drier the cooling rate becomes more rapid at higher levels. This drying process forms towering shower clouds, which produce heavy rainfall throughout most of the equatorial regions. This area of convective activity girdles the Earth, and is known as the 'Intertropical Convergence Zone' (ITCZ) (Fig. 2.5). Its position varies throughout the year, as it tends to follow the point where the Sun is directly overhead at noon. So, after mid-summer in the northern hemisphere (June 21) it spreads rain to the drier regions around 15°N reaching its most northerly position in August. It then follows the Sun southwards, but, after mid-summer in the southern hemisphere (December 21), the rainfall patterns, in general, spread far less below the equator reaching only around 5°S in February. This movement also means the region of easterly trade winds, to some extent, follow the Sun. The precise extent and position of the ITCZ is governed, however, by the position of the continents and oceans in the tropics (see Chapter 3).

Fig 2.5
A view of the Earth from a geostationary satellite showing the Intertropical Convergence Zone (ITCZ) girdling the globe in the vicinity of the equator which may, as here, include tropical storms (with permission of the Japanese Meteorological Agency.).

Another feature of the ITCZ is the degree of organisation in the convective activity. For the most part, the shower clouds do not group together into significant weather systems. But, in the right conditions, they can become more organised to form tropical storms, which can then develop into hurricanes and migrate to higher latitudes (see Section 3.9).

Because virtually all the moisture in the rising air of the Hadley Cell has been squeezed out of it in the form of rainfall, the descending air at 20°N and S is very dry. This means it warms more rapidly in descent than it did on ascent (at around 9·8°C per kilometre) and hence is hot and dry when it reaches the surface. So the continental regions around 20-30°N and 20-30°S are arid, and at sea the north-east and south-east trade winds normally feature dry, cloudless conditions. This simple model, first proposed over 300 years ago, works very well, but as we will see later, subtle changes in sea-surface temperatures (SSTs) across the equatorial oceans can lead to major changes in seasonal rainfall and wind patterns throughout the tropics.

Beyond the tropical regions the patterns become more complicated. This change reflects the fact that the tropics experience relatively small changes in temperature from place to place or from season to season. So broad generalisations can be made about the Hadley Cell, which are not applicable at higher latitudes. In these regions a different set of conditions apply. These are characterised by strong rotational motion, as the effects of the Earth's rotation increase with latitude, and substantial temperature contrasts. At low altitudes these physical effects combine to produce mobile, transient eddies (low pressure depressions) and more passive high pressure anticyclones.

At levels above about 3 kilometres (10,000 feet) these systems consist principally of wave patterns, generally moving from west to east. At any time the number of troughs and ridges may vary as will their position and amplitude. These patterns are known as *long waves* or *Rossby waves* (after the Swedish meteorologist Carl Gustav Rossby who first provided a physical explanation of their origin). The climatological mean of the flow has two major troughs at around 70°W and 150°W (Figs. 2.6 a and b). Their position is linked to the impact of the major mountain ranges on the flow, notably the Tibetan plateau and the Rocky Mountains, together with the heat sources such as ocean currents (in winter) and land masses (in summer). These waves are superimposed on a strong zonal current, the core of which is called the *jet stream*. They tend to fluctuate between a strong long wavelength pattern and a shorter meandering form (for physical reasons the number of waves girdling the globe is limited to three to six). These circulation patterns exert a fundamental influence on conditions in mid-latitudes because they guide the movement of weather systems (see Section 2.4).

The importance of low pressure systems in general circulation is the energy they transport polewards, especially in the winter half of the year. Their tracks are governed to a large extent by the distribution of the oceans and the continents. In addition, the effect of major mountain ranges is crucial. In the northern hemisphere winter, low pressure areas form over the warm oceans and high pressure areas over the cold continents (Fig. 2.7a). So the pattern is dominated by semi-permanent low pressure areas over the North Pacific and Iceland (in practice, these areas experience a sequence of low pressure systems passing through, but the average effect is for the average pressure to be low). Together with the subtropical high pressure systems, these systems sustain a strong westerly flow for much of the time.

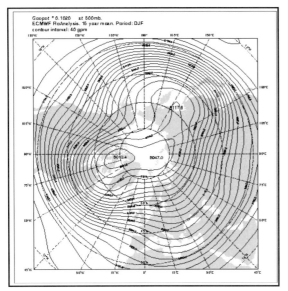

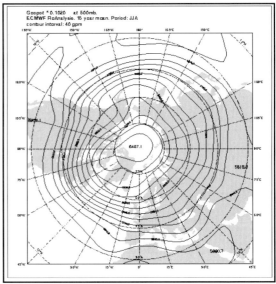

Fig 2.6 a & b
The mean height of the 500 mb pressure surface (in decametres) in (a) winter and (b) summer for the northern hemisphere, which exhibits a long wave structure. (With permission of ECMWF.)

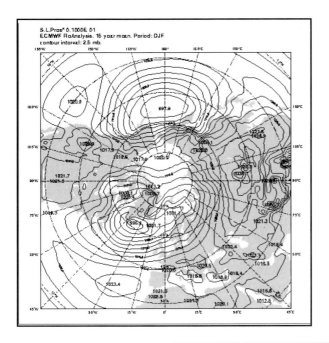

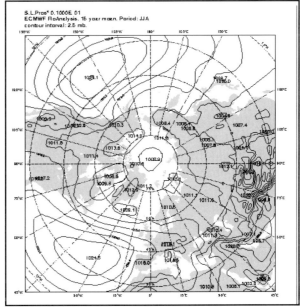

Fig 2.7 a & b
The mean surface pressure in (a) winter and (b) summer for the northern hemisphere. (With permission of ECMWF.)

23

The high pressure regions consist of one over Siberia, and one to the lee of the Rocky Mountains. The high over Siberia remains largely unmoved but is displaced far to the east as there is no barrier to Atlantic air moving in from the west, whilst the mountains to the south prevent any appreciable exchange with the Pacific and the continental areas to the south. Over North America the situation is more dynamic with incursions of arctic air alternating with either milder Pacific air or even warmer air from the Gulf of Mexico. In the summer, the pattern shows far less contrast (Fig. 2.7b) and is effectively reversed with high pressure over the relatively cool oceans and lower pressure over the warm continents.

In the Southern Hemisphere, the situation is much simpler. The combination of the near absence of continents in mid-latitudes and the almost symmetrical distribution of Antarctica about the South Pole reinforces the zonal circulation of the atmosphere. So both at sea level and aloft the flow is more uniform. The westerly winds are stronger and show much less variation from winter to summer. The pressure differences are less marked as the circumpolar track of the depressions as they move polewards effectively smears out the low pressure climatic feature into a latitudinal band around 60 to 65°S (Figs. 2.8a and b). In the upper atmosphere the patterns are also more symmetrical than in the northern hemisphere (Figs. 2.9a and b), and they fluctuate less markedly throughout the year.

Turning to the role of the oceans in the Earth's energy balance, roughly speaking they transport about the same amount of energy polewards as does the atmosphere. The broad pattern of ocean currents is well-known (Fig. 2.10). The forces driving these features are, however, complicated, and involve a series of interactions with the atmosphere. At the simplest level the oceans are driven by two processes. The first is wind stress. This imparts momentum to the ocean surface by forming waves. Broadly speaking many of the features of the surface ocean currents reflect the broad weather patterns described above. But the principal currents are more narrowly defined around the boundaries of the ocean basins.

The effect of the winds has a second important consequence. This is to create a mixed layer across the oceans of considerable depth, but with very little temperature difference from top to bottom. Beneath this mixed layer is the *thermocline* in which there is a rapid drop of temperature. The thickness of the mixed layer, and hence the depth of the thermocline depends on wind speed, thermal mixing where the surface waters are heated by the sun or altered by the passage of warmer or colder air, and by advection of warmer or colder water or the upwelling of cold water. All these processes can have a significant impact on sea surface temperatures and hence on the weather.

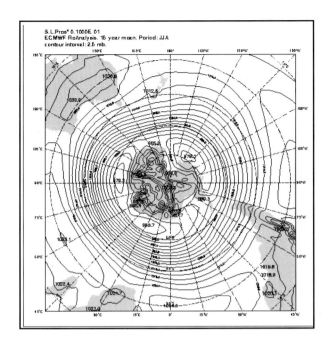

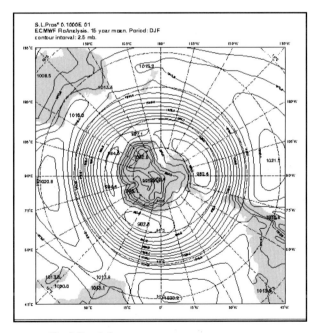

Fig 2.8 a & b
The mean surface pressure in (a) winter and (b) summer for the southern hemisphere. (With permission of ECMWF.)

25

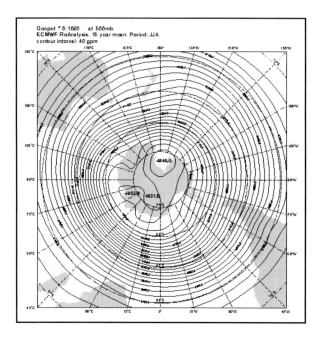

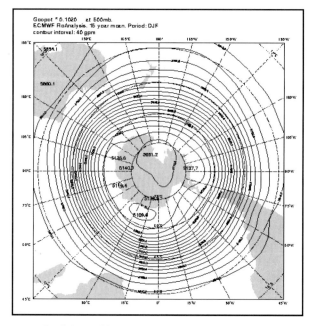

Fig 2.9a and b
The mean height of the 500mb pressure surface (in decametres) in (a) winter and (b)
summer for the southern hemisphere. (With permission of ECMWF.)

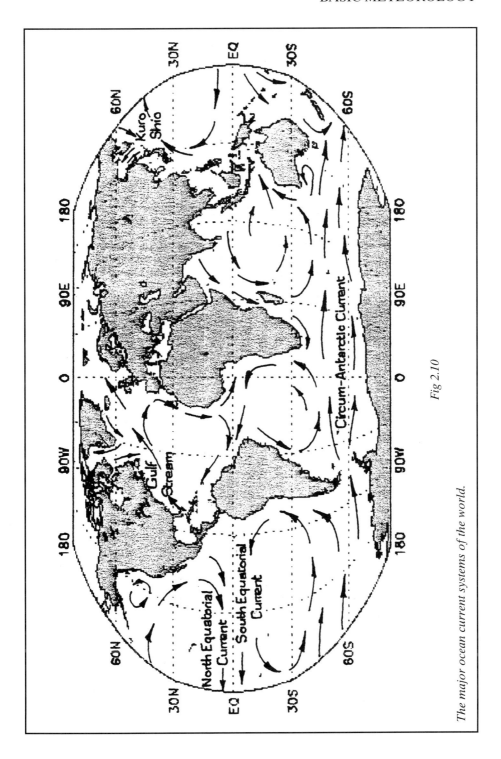

Fig 2.10

The major ocean current systems of the world.

The second process moving the oceans is thermohaline circulation. This is driven by changes in sea water density arising from variations in temperature and salinity and involves water sinking to great depths. The temperature depends on where the sea water has come from and on how much heat the oceans either pick up from or release to the atmosphere in sensible heat and evaporative loss. The salinity of a given body of water depends on the balance between losses through evaporation as opposed to gains from either rainfall or freshwater run-off from rivers and the melting of the ice sheets of Antarctica and Greenland plus the pack-ice of the polar oceans. In practice, there are very few regions where this process has a major impact. "Deep waters", defined as water that sinks to middle levels of the major oceans, are formed only in the North Atlantic, Greenland, Norwegian, Iceland and Labrador Seas. The world's "bottom waters" that form a colder denser layer below the deep waters are formed only in restricted regions of the Southern Oceans near the coast of Antarctica in the Weddell and Ross Seas.

Thermohaline circulation plays an important part in driving the ocean currents. The world-wide pattern of this circulation is known as the Great Ocean Conveyor (see Fig. 2.11). In terms of weather patterns, as warmer surface waters flow towards the sinking regions it is the heat they give to the atmosphere as they go that matters. But, it is the potential of the thermohaline process to fluctuate over time that is of greatest interest to maritime weather and climate, as this may be the cause of certain aspects of climate change (see Section 6.3).

In terms of the part played by various oceans in transporting heat polewards, the Atlantic carries rather more heat northwards than the Pacific (the Indian Ocean makes a negligible contribution because it extends so little distance northwards). In the Southern Hemisphere the Pacific transports roughly twice as much energy southwards as the Indian Ocean, while the Atlantic overall has a small negative effect carrying energy northwards across the equator.

2.3. Air masses

How the atmosphere transports energy in the form of dynamic weather systems is driven by the different properties of the bodies of air being mixed in each system. So, in understanding how these systems behave, it helps to start with a knowledge of where major bodies of air (*air masses*) come from and what this means in terms of their properties. This is particularly important at sea as the properties of the air are strongly influenced by interactions with the ocean it passes over.

TABLE 2.2
Classification of Air Masses

GEOGRAPHICAL

Air Mass	Symbol	Remarks
Arctic	A	Sources are the ice and snowfields of Greenland, the Arctic Ocean and Antarctica.
Polar (or subpolar)	P	Sources of maritime polar air (mP) are the Atlantic and Pacific Oceans north of 50°N and the Southern Ocean south of 50°S, while continental polar air (cP) spreads out from the anticyclonic regions over Siberia and Canada.
Tropical (or subtropical)	T	Sources of maritime tropical air (mT) are the high pressure areas over the oceans between 20 and 40°N and S, while continental tropical air spreads out of the hot dry continental areas of Africa, Asia, Australia and North and South America.
Equatorial	E	These air masses are principally maritime (mE) and are confined to the ITCZ apart from the notable excursion of the summer monsoon over the Indian subcontinent.

THERMODYNAMICAL

Cold	k	This air is colder than the underlying surface and so absorbs heat from below which generates convective activity as the air becomes "unstable".
Warm	w	This air is warmer than the underlying surface, and as it gives up heat the lower atmosphere cools, damping out any convective activity, and the air becomes "stable".

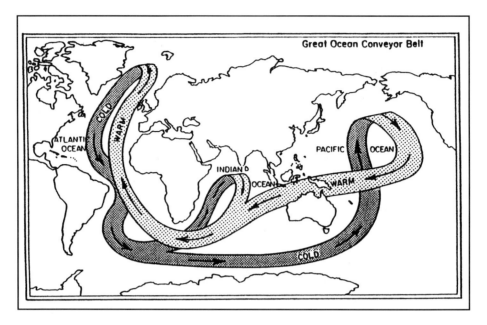

Fig 2.11

The Great Ocean Conveyor Belt - a schematic diagram depicting global thermohaline circulation.

Broadly speaking an air mass can be defined as "a widespread body of air that is approximately homogeneous in its horizontal extent, particularly with reference to temperature and moisture distribution; in addition, the vertical temperature and moisture distributions are approximately the same over its horizontal extent".[2] To meet these criteria the various parts of the air mass must be of common origin and have had the same "life history". So an air mass sitting over a warm ocean will acquire properties of temperature and humidity that are characteristic of the warm ocean area. Conversely, in winter air over the arctic snowfields will become very cold and dry both from contact with the surface and through the loss of heat radiation to space. As air masses move away from their source regions, they start to take on the properties of the surfaces they pass over and develop characteristic weather phenomena.

The classification of air masses was established by Professor Bergeron in the 1920s, in terms of two basic properties. The first was the *geographical* origin of the air mass, which was defined as one of four regions (see Table 2.2). The second was *thermodynamic*, and is simply a measure of whether the air mass is warmer or colder than the underlying surface and hence whether it is losing or gaining energy as it passes over the surface. This second property defines whether the air is becoming more or less stable with time (see Table 2.2).

The four primary types of air mass (Arctic, Polar, Tropical and Equatorial) are each divided into either continental (c) or maritime (m), depending on whether the source region of the air mass is a continental or oceanic area. Weather maps often label air masses according to this nomenclature. So, a polar air mass formed over a continent is designated "cP" and a polar air mass formed over an ocean is labelled "mP". A knowledge of the properties of air masses is an essential feature of understanding the Earth's climate (see Chapter 3), weather systems (see below) and also for interpreting weather forecasts (see Chapter 5).

2.4. Fronts, Depressions and Anticyclones

Many of the most interesting features of the weather occur where different air masses meet. Because contrasting air masses originate at high and low latitudes they clash in mid-latitudes, generating the dynamic weather systems that are an essential part of the global circulation system (see Section 2.2). The zone where these systems develop and move is generally known as the *polar front*. In the Northern Hemisphere the most important positions of this front for mariners are across the North Atlantic from Cape Hatteras to southwest Ireland and North Pacific from Japan to Oregon. This front moves southwards in winter with the Atlantic front extending into the Gulf of Mexico and the Pacific front moving down into the China Sea. Both fronts retreat northwards in summer. In addition, in winter an important frontal zone develops along the northern edge of the Mediterranean. In the Southern Hemisphere the polar front has a simpler circumpolar structure and moves less throughout the year.

In the vicinity of the polar fronts a succession of depressions (low pressure systems) are formed. Each depression is fuelled by the organised mixing of cold, dry arctic or polar air with warm, moist air from low latitudes. The standard model of a mid-latitude depression was first proposed in 1919 by the Norwegian meteorologist, Vilhelm Bjerknes. The important feature of the model is that, throughout the lifetime of the depression, the separation between the two air masses is normally sharp (Fig. 2.12). Along this front temperature and wind change suddenly, and in most cases the state of the sky also changes. Sections of the front are named according to their motion: a *warm front* is a front along which warm air displaces cold air; a *cold front* has the reverse characteristic. A cross-section of a typical depression to the south of its centre shows wedges of cold air under warm air. To the north of its centre the cold air forms a trough filled with warm air. The overall effect of the depression is to raise the warm air creating extensive layers of cloud and precipitation in the region of the warm and cold fronts.

The life cycle of a depression, as proposed by Bjerknes, starts with a quasistationary front between a cold and warm air mass with a shearing motion between them. This develops into a wave which eventually winds up as a well-developed depression. In this process the cold front catches up with the warm front using up the wedge of warm air between them to drive the depression through its development. Where the cold front overtakes the warm front the system is said to 'occlude' and the resulting *occluded front* has no temperature change at ground level but may be marked by heavy precipitation. In its later stages the depression runs out of energy and begins to stagnate, but its trailing cold front can become the site for the next *secondary* depression. In this way a family of several depressions can follow one another along similar tracks. While these tracks are along the polar front, the position of this zone can vary substantially from time to time.

Most meteorological textbooks present the family of depressions as a rather orderly process. Rather like a troop of elephants trunk-to-tail in a circus ring, they are shown girdling the globe. In real life it is a different story. Some secondary depressions are pale shadows of their forebears, while others rapidly grow into monsters which dominate the circulation of a significant part of the hemisphere. The factors that affect whether minor disturbances do or do not grow are central to weather forecasting which is discussed in Chapters 4 and 5. So, for the moment, the standard model of strings of depressions transferring energy poleward is useful and reinforces the essential continuity of the model of depressions in which each eddy is a natural product of its predecessor.

The Bjerknes model cannot explain why the frontal zones in most depressions are so sharp. Indeed, this remains an area of scientific research. But for mariners the essential requirement is to recognise the changes associated with the passage of a front as these can exert considerable influence on the safe and efficient operation of shipping. So these changes, which are described briefly in Fig. 2.12, will be explored in more detail in both Chapters 3 and 5.

Anticyclones lack the organised dynamics of depressions. Often they appear as sluggish, passive systems which fill the spaces between their vigorous cyclonic counterparts. In part, this reflects the fact that many

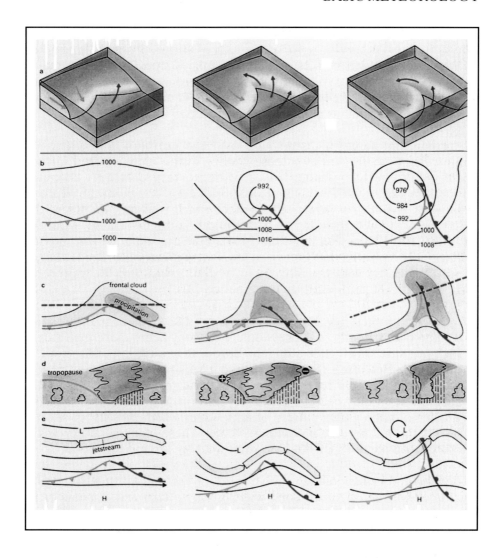

Fig 2.12

The stages of development of a frontal depression: (a) the view from above shows how the wave on the frontal surface grows and finally occludes over a period of days; (b) shows the warm and cold fronts at the surface and the associated atmospheric pressure distribution; (c) shows the distribution of frontal clouds and precipitation; (d) indicates the cloud development and precipitation in relation to the vertical structure of the frontal zones along the cross sections indicated in (c); and (e) shows the position of the fronts at the ground and the flow aloft including the jet stream axis which is typically associated with a frontal depression. The middle diagram of (d) shows the jet stream entering the page (+) and leaving the page (−). (From Burroughs, 1991, with permission of Cambridge University Press.)

anticyclones are the source of the air masses which fuel depressions in mid-latitudes. As such, in spite of the lack of uniformity in their patterns of formation, growth and decay, and irregular shape, they play a major role in the dynamics of weather patterns. The connection with air masses is also recognised in the broad categorisation of anticyclones into 'cold' or 'polar' continental highs and 'warm' or 'dynamic' highs.

Polar continental highs develop over the northern land masses in winter. They are created by intense cooling of the snow-covered surface which gives rise to a shallow layer of dense very cold air. Dynamic anticyclones are caused by large-scale subsidence throughout the depth of the lowest 10 km or so of the atmosphere which is usually known as the *troposphere*. They include the subtropical highs associated with the subsidence of the descending limb of the Hadley cell, and slow-moving mid-latitude highs that are known as *blocking anticyclones*. The latter are closely associated with the movement of depressions and the pattern of long waves in the mid-latitudes (see Section 2.2).

At the simplest level dynamic highs are the areas of high pressure that punctuate a family of depressions. They may be no more than ridges of high pressure which glide along the border of a much larger subtropical anticyclone. But often they can assume a separate identity and play a much more influential role in affecting weather in the mid-latitudes. In particular, when they become stationary and form an integral part of the establishment of a global pattern of long waves which persists for several weeks, then they become a dominant force. These blocking anticyclones play a central part in spells of abnormal weather around the world.

As noted in Section 2.2, the strength of the mid-latitude westerlies fluctuates between a strong long wavelength pattern and a meandering short wavelength pattern. The latter is frequently associated with the establishment of blocking conditions. These are defined by the upper-level westerlies being split into two well-defined branches which extend over at least 45° of longitude. To constitute a block, these cells of high pressure, which are sandwiched between normal zonal westerly flow, must last at least 10 days. This condition is frequently met, as once blocks form they are relatively stable persisting on average for 12-16 days, although they can last much longer. In the Northern Hemisphere their position is influenced by the distribution of the continents and mountain ranges. They most frequently occur close to the Greenwich meridian and in the eastern Pacific. Atlantic blocks are approximately twice as common as the Pacific variety. In the Southern Hemisphere blocking is less common - a fact that may be related to the relative absence of land masses in the southern mid-latitudes.

At the more detailed level, it is important to recognise how the behaviour of anticyclones over the oceans clearly shows up the stability of the air masses involved. Plunges of polar or arctic air behind cold fronts over the oceans rapidly become highly unstable, with the generation of a characteristic pattern of intense convective activity. This behaviour shows up in many satellite pictures as a mottled mass of shower clouds. Conversely, warm anticyclones spreading to higher latitudes over colder oceans produce the standard combination of fog and decks of low stratus cloud. The consequences of atmospheric stability in a wider variety of weather situations will be discussed in Chapters 3 and 5.

2.5. Wind and Waves

The term wind stress (see Section 2.2) was introduced in discussing the formation of ocean currents. Its more immediate consequence, when associated with any weather system, is the generation of waves. Although many aspects of understanding and forecasting the state of the sea are matters of seamanship, a basic knowledge of the factors that control the height (the vertical distance between successive peaks and troughs in a wave-train) and wavelength (the horizontal distance between successive wave crests) is an essential part of maritime weather and climate.

The first thing to get straight is what we are talking about when considering wind and waves, as both are highly variable quantities. This is a matter of definition. In the case of the wind, most of the time we will use the mean speed at a height of 10 metres (33 feet), although from time to time short term gusts will be quoted to provide an indication of the damaging conditions which can occur in severe storms. The mean wind speed at 10 metres over the open sea is approximately the speed which would be estimated by an experienced mariner from a visual assessment of the state of the sea. Wind speeds in most marine weather forecasts are for a height of 10 metres above the sea surface. Over the open sea the 10-metre wind speed is about two-thirds of the wind speed at the top of the boundary layer (about 900 metres above sea level). The value of using the figures at 10 metres is they relate directly to the values produced by weather forecasts, and also they tie in to the climatological statistics obtained both from direct observations and also to figures obtained from satellite measurements (see Chapter 3).

When it comes to waves, the most widely used statistic is the *significant wave height*. This is the average height of the highest third of the waves in any sea state. This definition of wave height was first developed during World War II and reflects the fact that observers are inevitably influenced by the bigger waves. This meant visual estimates of average wave height were found to equate to the significant wave height, whereas the average height was found to be just under two thirds of the observer's estimate. Furthermore, in all aspects of maritime operations and engineering it is the big waves that matter and so working with a statistic which focuses on the bigger brutes is a sensible approach.

The next thing to concentrate on is the fact that the current sea state at any point on the ocean is the product of current weather conditions and events in the recent past. In short, it takes time for the wind to build up waves and these waves take a long time to die away once the wind has abated. The waves also move beyond the limits of the geographical area in which they were generated. In addition, the height of the waves depends on the time the wind has been blowing at a given speed (Fig. 2.13), and the fetch to windward (the distance across the ocean over which the wind has been blowing) (Fig. 2.14).[3] Also, as waves move into shallower water initially they become higher and their wavelength shortens so they become steeper. Here we will concentrate on what is known about wave statistics in deep water and for unlimited fetch. These figures do, however, have to be interpreted by mariners when used in combination with observations of local conditions and weather forecasts (see Chapter 5).

As Figures 2.13 and 2.14 show, the generation of waves by wind stress takes time and space. To understand the behaviour presented in these diagrams it helps to know a little about the physical properties of waves. The first thing to appreciate is that the speed of a simple wave is directly proportional to its period (i.e. the time for successive peaks of a wave to pass a fixed point). The simple relationship is that the velocity of the wave (in knots) is just over three times the period (in seconds). So a wave with a period of 10 seconds travels at 30 knots, while a wave with a 20-second period travels at about 60 knots. At the same time the wavelength of the wave (the distance between successive peaks) increases as the square of the period, so a 10-second wave has a wavelength of just over 150 metres (500 feet) and a 20-second wave has a wavelength of over 600 metres (2000 feet). So when a steady wind of, say, 30 knots blows across a large expanse of ocean for a sustained period the initial waves are small, slow-moving and of short period. As they pick up more energy from wind they accelerate and their period increases. But as their speed approaches the wind speed they cease to get any extra impetus from the wind and so their period has an upper limit of around 10 seconds.

The size of any specific wave is limited by its steepness. Beyond a certain gradient the wave will start to break forming *white caps or white horses*. Well away from the shore, this limit is reached when the height of the wave is about one thirteenth of the wavelength. This means a 10-

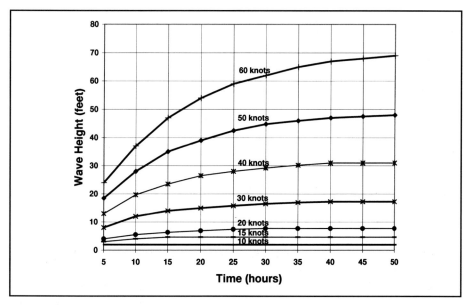

Fig. 2.13.
The height of waves at any given wind speed depends on how long they continue to blow.

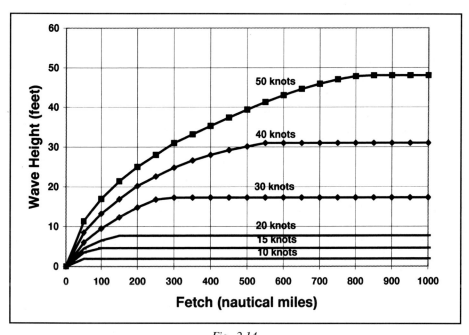

Fig. 2.14.
The height of waves for any given wind speed depends on the extent of any expanse of water (fetch) they blow over.

second wave with a wavelength of 150 metres (500 feet) could reach a height of about 12 metres (38 feet) before breaking, while a 20-second wave could, in principle, reach a colossal height of 46 metres (150 feet) if sustained long enough by a 60-knot wind. So, while the breaking of waves acts as a real physical limitation for shorter-period waves (e.g. up to about 8 seconds) for longer-period waves this limit has no practical effect. As waves approach the shore and enter shallower water their properties change. When and how they break is an entirely different matter depending, as it does, on local topography.

In practice, the sea state is more complicated, being made up of waves generated by the current winds and any swell produced by earlier weather, in distant wind systems. At any point the significant wave height (H_{sig}) is given by the expression[4]

$$H_{sig} = (H^2_{swell} + H^2_{wave})^{1/2}$$

where H_{swell} is the height of the swell and H_{wave} is the significant height of the waves generated by the local windfield. Because of the variability of wind speed both at any given point and across any weather system, there are always a range of wavelengths and heights in the sea. Furthermore the wave height is not a simple sum of the various waves in the resulting sea state but is a complex non-linear combination of all the disturbances at the time. This has a variety of consequences, the most obvious of which is that the wavelength of the sea state for any given period tends to be rather shorter than the simple wave formula. Therefore, because of the non-linear nature of the relationship between the significant wave height and the swell and windfield components, the instantaneous wave height may be substantially different from the value given by this simple equation. This is especially true where wind and swell directions are significantly different. In these circumstances the wave shape can become pyramidal. If, at the same time, the wind speed is rising rapidly then the sea state becomes even more complicated, and it has to be a matter of seamanship to interpret the condition.

More generally, the fact that usually the wind is either rising and generating waves or declining and being overtaken by the waves means that the sea state is rarely ever in equilibrium with the wind regime. As a practical observation, the wave speed tends to be greater than the wind speed below 25 knots, and less than it for higher values. Any underlying swell will be travelling faster than the locally generated waves. Once it leaves the track of the parent storm, where energy is dissipated in generating white caps, it spreads across the ocean with little decrease in height following a Great Circle path. Swell can travel great distances.

Waves generated by storms in the vicinity of Cape Horn with a period of 20 seconds (i.e. the product of sustained winds well in excess of 60 knots) have been observed in Cornwall 10 000 to 12 000 kilometres away, having taken 9 days to cover the distance.[5]

Long-lived swell can have a major impact as it enters shallower water. For instance, the arrival of swell in exposed areas west of the Shetland Islands is of considerable concern to offshore oil operations (see Section 4.4). There is also evidence of swell from storms in the western North Atlantic (see Section 3.6) generating huge "shoaling" waves as they move inside the hundred-fathom line of the Bay of Biscay approaching the north-west coast of Spain. These are not only a hazard to shipping but can also do great damage to low-lying land along the coastal areas of western Europe. A good example of this hazard occurred in the English Channel on 13 February 1979.[6] A deep depression (952 mb) south of Newfoundland two to three days earlier moving at the same speed as the waves it was generating (30 knots) led to a swell period of 18-20 seconds moving out across the North Atlantic. At weather ship LIMA (57°N, 20°W) and Data Buoy 1 in the Western Approaches a swell with a period of 18 seconds and a significant wave height of seven metres (23 feet) was observed at 0000 hours on 13 February, while around the same time observations south of Lisbon recorded a wave of 17·2 metres (56 feet) with a period of 20 seconds. Although the wave heights were not that exceptional, the combination of the very long wavelength and a moderate tidal surge (see Section 4.6) resulted in severe flooding along the coast of the English Channel. The distant effects of deep depressions also play a major part in the wave climate of the tropics, where the sea state is dominated by swell generated at higher latitudes.

Just how big waves can get is the subject of nautical legend. Perhaps the biggest reported wave is a monster claimed on the basis of visual estimates by observers on the *USS Ramapo* in the North Pacific in 1933 was reckoned to be 112 feet from crest to trough,[7] and there have been many other reports of waves between 80 and 100 feet high. But these reports were based on visual estimates. Reliable physical measurements were, until recently, much fewer and far between. As noted in Section 1.2 automatic buoys and satellite observations are now providing new insights into how big waves can get and are tending to confirm earlier reports of huge waves being a feature of severe storms.

Before embarking on what extremes have been observed in recent years, it is illuminating to consider what, in theory, could happen. In any sea there are a large number of wave components, each with their own period and height, travelling at slightly different, but constant speeds. As

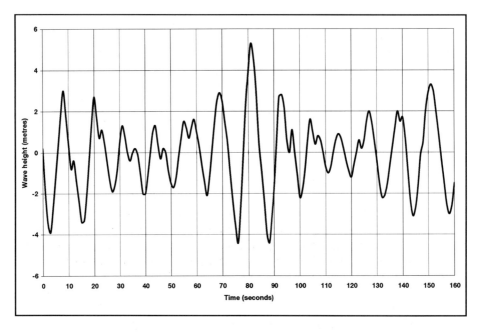

Fig 2.15
An example of how the height of waves passing any given point varies over time with considerable variation in the height of successive peaks and troughs.

the components continually get into and out of step with each other they produce groups of high waves followed by relatively quiet water (Fig. 2.16). This can be a relatively orderly process because the principal disturbances are centred about a wavelength which reflects the mean wind that has generated the sea state, and so their wavelengths do not differ by a large amount, say, by around plus or minus 20 per cent. This means that they add or subtract from one another over some five to ten wavelengths producing big waves at approximately this spacing - the *Seventh Wave* of nautical legend - with smaller waves in between.

In some sea states things are more complicated, especially when the wind-driven waves and any underlying swell, from one or more distant sources, are moving in different directions. In these circumstances every now and then, by chance, a large number of the components will get into step at the same place and form an exceptionally high wave. This is a transient phenomenon which disperses within a minute or two as the components move out of phase. This process means that there is theoretically no limit as to how the random jumble of waves on the sea can come together to produce what is popularly known as a *freak wave*.

This term is, however, not helpful as it implies something un-natural, whereas, what we are concerned about is a set of statistical rules. These imply that one wave in ten will be nearly 30 per cent above the significant wave height, and only one in 500 will be nearly twice the average height of the highest third of the waves. For practical purposes this means that the maximum wave height is likely to be about twice the significant wave height, but there is a roughly one in a hundred thousand chance of encountering a wave that is two and a half times the significant wave height (often defined as an *extreme wave*).

The distribution of significant wave height at any point over time is more difficult to describe. Any analysis must reflect not only seasonal variations, but also the occurrence of major storms (e.g. tropical cyclones or deep extratropical depressions), which can be regarded as a subset of the range of events likely in any one place. Rather than seek to describe possible arcane statistical theories, which may be able to reflect the spread of conditions, it is better to take the practical view that wherever exceptionally severe but rare events are known to occur the distribution of wave heights will have a very long 'tail' at the highest values. They are not 'freak' events. They are part of the overall climate of the area, even though they have a very low frequency of occurrence. The climate of these events is not well known and is not well documented in any published literature. So where good examples of these extremes have been recorded they will be highlighted in this book to provide an idea of just what maritime operations may have to take in their stride.

The other form of extreme that may form in rapidly developing weather situations is exceptionally choppy seas. These occur when wind speeds increase dramatically either with intense thunderstorm and squall lines, or with *explosive cyclogenesis*. The latter occur when newly formed extratropical depressions deepen suddenly. Any low which deepens more than 24 mb within 24 hours is often designated a *meteorological bomb,* while more extreme cases (e.g. a small depression off the Chesapeake Bay with central pressure of 996 mb at 0000 hours on 4 January 1989 had by 24 hours later deepened to a 936 mb brute as it moved to south west of Sable Island - a drop of 60 mb) are sometimes termed *ultrabombs*. In these circumstances, the winds increase so rapidly that the wave heights grow faster than the wavelength leading to steep seas. Extreme conditions like this can produce significant wave heights of 8 metres (26 feet) or so with a dominant period of around 10 seconds (i.e. a mean wavelength of around 150 metres [500 feet]),[8] but with some waves being substantially bigger: a truly awful sea state.

The final aspect of how wind and water can interact is the change in

41

sea level in well-developed depressions and tropical storms. The combination of the organised circulation of the storm and its motion can build up a substantial rise in sea level and in certain situations this develops into a *storm surge* which can do great damage when it hits land. In the case of extratropical depressions these are most dangerous where the storm surge is funnelled into narrowing and shallower waters such as the southern North Sea (see Section 4.6). With tropical storms and hurricanes the increasing swell is a sign of the approaching storm while the storm surge is often the most damaging element of the storm, as we will now see.

2.6. Tropical Depressions, Storms and Hurricanes

As already noted in Section 2.2, the circulation in the tropics is relatively simple because of the small changes in temperature from place to place, or season to season. Nonetheless, they can spawn the most formidable storms, which sweep outward to higher latitudes and are an important part of the general circulation of the atmosphere in late summer and early autumn. Starting initially as *tropical depressions* these systems can

Fig 2.16
A cross-section of a hurricane shows how the circulation at low level pulls in moist warm air which rises up in the centre of the storm and spreads out at high levels (with permission of NASA).

Fig 2.17

The most intense typhoon (Tip on 12 October 1979) ever measured in the Pacific. This infra-red image shows the extraordinary symmetry of intense tropical storms. At the time the central pressure in the clearly defined 'eye' of the storm was 870 mb (with permission of the Japanese Meteorological Agency).

develop via *tropical storms* to full-blown hurricanes (see Table 2.3). These storms have their genesis over the tropical oceans in weak wave features and minor depressions occurring along the edge of the ITCZ (see Section 2.2). These *easterly waves* are a common feature of tropical weather and are similar to fronts at higher latitudes, and frequently produce squally weather and heavy rain. Many of these features live out their existence as no more than weakly organised convective activity, but some of them develop into major systems. What causes one disturbance to come to nothing and another to grow explosively is still the subject of research which will be discussed in detail in Chapter 5 in terms of forecasting the development and course of hurricanes.

Once the conditions are right for hurricane development the storm grows rapidly and often follows a standard pattern. As the pressure starts to fall rapidly at the centre of the disturbance, the winds rise in a tight band of some 30 to 60 kilometres radius and the clouds form into a remarkably circular pattern about a central eye (see Figs. 2.16 and 2.17). The circular cloud system then expands as the storm matures. The maximum wind speed around a storm is highly correlated to the degree of organisation and the diameter of the circular area.

As the storm grows, it moves westward in the trade winds, usually in the belt 8 to 15° from the equator (Fig. 2.18), and starts to move towards higher latitudes. The maturing storm then expands while the central pressure stops falling. The speed at which it travels and the course it follows depends on the underlying surface conditions and the winds in the upper troposphere at levels around 12 km (40,000 feet). The climatology of the erratic nature of hurricane movements is reviewed in Section 3.9, and progress on forecasting their movements in Section 4.8. What is clear is that a fundamental ingredient in their development is sea surface temperature (SST). Hurricanes only develop over water with a temperature above 27°C and their development and intensity is related to the pattern of SSTs. Their course is, however, more influenced by steering winds in the upper atmosphere. So the combination of being fuelled by the oceans from below and steered by high level winds makes them a major forecasting challenge.

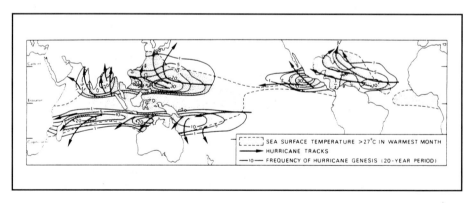

Fig 2.18

Frequency of hurricane genesis (numbered isopleths) for a 20-year period. The principal hurricane tracks and the areas of sea surface having temperatures greater than 27°C in the warmest month are also shown (from Barry & Chorley, 1986, Fig. 6.9).

TABLE 2.3
Properties of Tropical Depressions, Storms and Hurricanes

Type	Category[1]	Central Pressure (mb)	Maximum Sustained Wind speed (knots)	Height of Storm Surge (feet)	Damage[2]
Tropical Depression			<35		
Tropical Storm			35 -64	<4	
Hurricane	1	>980	65 -83	4 -5	Minimal
Hurricane	2	965-979	84 -95	6 -8	Moderate
Hurricane	3	945-964	96 -113	9 -12	Extensive
Hurricane	4	920-944	114 -135	13 -18	Extreme
Hurricane	5	<920	>134	>18	Catastrophic

FOOTNOTES

1.   The categorisation of hurricanes is defined as the Saffir-Simpson scale after the meteorologists who developed it. Tropical depressions and storms are not the subject of categorisation and classified solely on windspeed as this is the only distinguishing measure of their behaviour - they may or may not graduate to hurricane status or simply fizzle out.

2.   The damage categories relate almost entirely to the impact of storms on the shoreline and further inland, but can include losses at sea.

This mention of upper atmosphere winds raises another aspect of behaviour at the high levels which rarely intrudes into day-to-day meteorology, but which has implications for hurricane forecasting. This is the periodic reversal of the winds in the lower stratosphere at levels between around 20 and 30 km (65,000 and 100,000 feet) over equatorial regions. These winds go through a cycle every 27 months or so from being strongly easterly to nearly as strongly westerly. This behaviour is known as the Quasi-Biennial Oscillation (QBO) and appears to have weak echoes in many features of the weather at lower levels, including the incidence of hurricane activity, in the Atlantic (see Section 4.7). It is also the only clearly defined cycle in the weather longer than the dominant annual cycle (see Section 6.3).

Some tropical cyclones can grow to a radius of more than 300 kilometres before they start to decay. The declining phase is accelerated by either passing over land or colder water. By this stage the hurricane has reached higher latitudes and is being pushed eastward by the mid-latitude westerlies. As a measure of the threat they continue to pose to shipping, the encounter between the *Queen Elizabeth II* and Hurricane *Luis* on 11 September 1995 at around 43° 30' N, 54° 09' W, is a good illustration.[9] Maximum winds of 120-130 knots with seas up to 24-25 m (79 feet), with an extreme wave of 29 m (95 feet) forced the liner to reduce speed to 4.5 knots (the minimum for steerage) and caused considerable damage to deck plating, superstructure and windows.

At higher latitudes, hurricanes can have one further trick up their sleeve. This is that they may sometimes pick up a new lease of life with great rapidity by either becoming reinvigorated or by transferring energy to a nearby system to generate an explosive new extratropical depression. Developments of this type can represent a real threat to shipping in the northern hemisphere and a major challenge to weather forecasters (see Section 5.4). A good example of this type of development was hurricane *Lili* in October 1996. This storm crossed Cuba and moved south of Bermuda between 18 and 20 October. By the 22nd it was at 33°N 54°W and the Miami Tropical Cyclone Prediction Center forecast it would decay within 24 hours, and stopped issuing advisories. Instead it started to revitalise and move towards Europe. On 25 October, it was still classified as a hurricane and during 24 hours it spawned an intense low (Low T) to its north which deepened 30 mb over 24 hours and reached Scotland on the 27th. In the meantime, *Lili* became an extratropical storm (Low L) which reached western Ireland on the 27th and with a central pressure 975 mb ran across Wales and England on 28 October bringing wind of up to 70 knots. These developments were not well forecast and provide a good example of how hurricanes moving to higher latitudes can produce a train of complicated developments. In particular, how energy was transferred to ahead of *Lili* to Low T to enable this new system to develop so rapidly, remains an unsolved issue.

At sea the greatest threat to safety from hurricanes is the high winds and waves. In the northern hemisphere the highest waves are generated in the direction of the hurricane's travel on its right hand side. This is because the winds from the right rear quadrant of the storm blow in the direction of the hurricane's movement and so the waves propagate under the influence of winds having relatively little change of direction for longer than in any other part of the storm. These waves and strong swell travel ahead of the storm and provide warning of its approach. As the storm comes near to the shore the waves pile up ahead of the hurricane forming storm surge in the region of the greatest winds. This wall of water can do the most damage when hurricanes strike the shoreline, especially if it coincides with a high tide.

The classification of tropical cyclones often relies on the Saffir-Simpson scale (Table 2.3), which links central pressure, wind speed and storm surge height. Strictly speaking this scale applies only to hurricanes in the Atlantic basin, as the properties of tropical cyclones in other parts of the world are slightly different (e.g. the central pressure in storms in the north-west Pacific tend to be about 6 mb lower for comparable wind speeds - see Table 2.4). Furthermore, because many of these storms develop in remote areas where there are few, if any, measurements, they are often classified in terms of their satellite images. An important feature of this analysis is the degree of symmetry exhibited by the storm: the more symmetric it is, the more intense it is (see Fig. 2.17). This classification is known as the Dvorak scale[10] and is given in Table 2.4. This scale is widely used by tropical analysis centres around the world and digitised methods of pattern recognition have been developed to remove the subjective element of measuring the intensity of storms,[11] and these continue to be refined to improve the use of satellite images in forecasting work.[12]

TABLE 2.4.
The Dvorak Scale

Current Intensity (Number)	Wind speed (knot)	Mean Sea Level Pressure (Atlantic - mb)	Mean Sea Level Pressure (Northwest Pacific - mb)
1	25	—	—
2	30	1009	1005
3	45	1000	994
4	65	987	981
5	90	970	964
6	115	948	942
7	140	921	915
8	170	890	884

2.7. El Niño Southern Oscillation

So far this presentation of basic meteorology has concentrated on circulation processes that transport energy from low latitudes towards the poles, and their associated weather systems. There is, however, one major aspect of tropical meteorology which affects the distribution of weather patterns across the tropics which is best introduced here. This is an irregular fluctuation of SSTs across the equatorial Pacific which exerts a profound influence over tropical weather. Many of these will be considered in terms of their impact on weather forecasting (see Chapter 4) and climate change (see Chapter 6).

Peruvian fisherman have known for centuries that warm water tended to spread southwards from the equator at the end of each year. This water capped cold upwelling nutrient-rich water which sustained abundant stocks of fish. In some years these warm events were much stronger and had a catastrophic impact on the fishing. Because of their timing they were linked with the Nativity and called El Niño (the little boy in Spanish). It is now known that the interannual variation of these events is linked to much wider atmospheric-ocean interactions.

The other part of this tropical system was first identified in the 1920s by Sir Gilbert Walker, when he was Director General of the Indian Meteorological Service. He noted: 'When pressure is high in the Pacific Ocean, it tends to be low in the Indian Ocean from Africa to Australia'. He named this pattern the Southern Oscillation, and it is now recognised as being directly linked to the El Niño. For this reason, they are usually considered together as the El Niño Southern Oscillation (ENSO).

The essential features of the ENSO are shown in Fig. 2.19. During normal conditions atmospheric pressure is higher in the eastern and central Pacific and lower in the west, and the equatorial winds blow from South America towards New Guinea. This drives the ocean currents westwards which produces lower sea levels in the east than in the west with cold water upwelling along the coast of South America bringing the thermocline to the surface. This means SSTs are low close to the equator (typically 20 to 24°C) and extend westwards a in long cool tongue towards the International Dateline which suppresses convection. In the west of the Pacific the SSTs are high (typically 27 to 28°C) and this generates a region of strong convection. This atmospheric pattern established a convective loop which reinforces the climatic pattern.

When the ENSO reverses the surface winds in the western tropical Pacific flow westward and the eastern and central tropical Pacific becomes warmer, wetter and the atmospheric pressure drops. Warmer

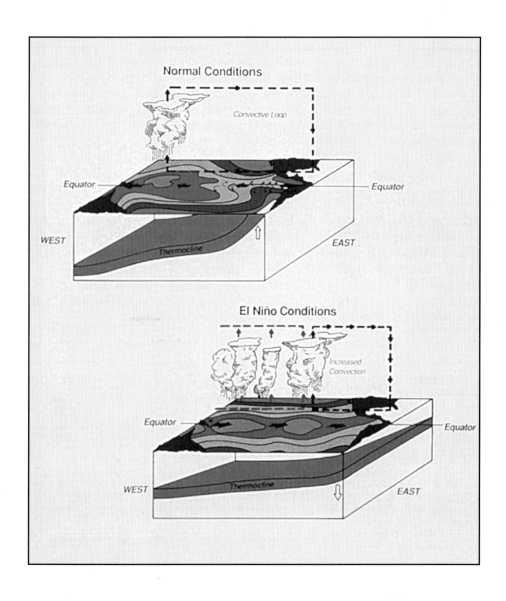

Fig 2.19
Schematic illustration of the differences in the tropical climate between normal and El Niño conditions. For the latter, the thermocline becomes less tilted, sea surface temperatures increase in the eastern Pacific and the regions of the central and eastern Pacific see increase convection. (From IPCC, 1995. Fig. 4.7.)

water extends eastwards and the thermocline is depressed. The area of strong convection expands towards South America and this establishes an atmospheric circulation which tends to reinforce the El Niño conditions.

For weather forecasters and climatologists the challenge of the ENSO is to explain why, once established either normal or El Niño conditions do not remain in place interminably. Models of how every few years the ENSO can swing back and forth in a quasi-cyclic manner are the key both to understanding the processes at work and producing better forecasts of how the ENSO will develop over timescales of three to twelve months ahead.

For mariners the changes occurring across the equatorial Pacific are of limited direct interest as this is not an area covered by busy shipping lanes, and the changes are, in themselves, not major threats to shipping. What really matters about the ENSO is the part it plays in the fluctuations of extreme weather events around the world and the tropics in particular (e.g. the incidence of hurricanes in the North Atlantic and other tropical oceans, and the strength of the summer monsoon in the Bay of Bengal and the Arabian Sea). So both the ability to forecast changes in the ENSO and to use this to predict extreme weather events around the world will be considered in more detail in Chapters 4 and 6.

FOOTNOTES

[1] Many of the features of the Earth's radiation balance are covered in more detail in the standard texts on meteorology and climatology listed in the Bibliography, so any reader who wishes to explore these concepts in greater physical depth is advised to turn to these books.
[2] Definition provided by the American Meteorological Society, Glossary of Meteorology, (1959).
[3] Draper, (1966).
[4] Young & Holland (1996).
[5] Barber & Ursell (1947).
[6] Draper & Bownass (1983).
[7] Nickerson (1993).
[8] Dorr (1990).
[9] The Marine Observer, (1996). **66**, 134-137.
[10] Dvorak (1975).
[11] Dvorak (1984).
[12] Velden, Olander & Zehr (1998).

CHAPTER 3

CLIMATOLOGY

As the Spanish proverb says, 'He, who would bring
home the wealth of the Indies, must carry the wealth
of the Indies with him.' So it is in travelling; a man
must carry knowledge with him, if he would bring
home knowledge.

Samuel Johnson, 1709-1784

The climatology of the oceans sounds a daunting prospect. The thought
of reams of statistics about climatic conditions around the world is not an
appealing one to many mariners. That is as may be, but the fact of the
matter is that without a good knowledge of what normally happens to the
weather in any part of the world, it is not possible to understand fully the
guidance being given by local meteorological services. Moreover, it is not
just a matter of knowing what the average conditions are likely to be, but
also appreciating what extremes may also occur because many extreme
and dangerous conditions are part and parcel of the local weather. So
without the grounding of climatological knowledge, it is not possible to
make effective use of forecasts and other meteorological advice.

In concentrating on those elements of the climate that constitute the
greatest hazards to maritime activities, our analysis will not explore
certain areas in any detail. In particular, the distribution of temperatures
across the oceans, and its seasonal variations will be passed over. This
information can be found in standard climatological texts (see

Bibliography). Where temperature variations are important in terms of either the impact of local weather or the contrasts between sea and air (e.g. the risks of poor visibility and fog) and the related question of being abnormally high in winter, or well below normal at any time of year which can be a sign of impending squalls or sudden storms, or the consequence of longer term variations in sea surface temperatures (e.g. as part of climate change or in defining fishing stocks), these aspects will be highlighted where appropriate. In short, our objective is to identify those elements of the climate which are of particular relevance to maritime operations, whatever type of vessel or installation you are responsible for and whatever activities you are conducting.

3.1. Sources of Data

The scale of climatological data for the oceans is changing dramatically. This is because there are two new sources of data to draw upon. The first is satellites, and the second is automated buoys. Various satellites (e.g. GEOSAT, ERS 1 and 2, and Topex/Poseidon) have begun to provide continuous coverage of the behaviour of oceans. Prior to this mariners had to rely on the distillation of shipboard measurements.[1] While these provide a broad picture of the conditions mariners are likely to experience on the shipping routes around the world they are far from complete. Moreover, in spite of their limitations, they still represent a huge body of material which could not possibly be covered here. So, we will focus on two things. First, we will consider those areas where new measurements are providing important new insights into conditions at sea. Secondly, we will concentrate on what constitutes the greatest hazards to shipping, notably, wind, waves, storms and sea-ice. This requires the combination of information on which elements of the weather are most important from place to place, what can normally be expected, and then concentrate on just how extreme conditions can become as an indication of what can be expected in planning operations in different parts of the world.

In providing this selective climatology, we draw on the various marine atlases that have been published by the UK Admiralty and the US Navy, some of which are now available in computer compatible form[2] and the first fruits of the satellite era (e.g. the data obtained from the altimeter on GEOSAT between November 1986 and January 1990).[3] Although the satellite data cover only a short period, where they are capable of providing accurate measurements, they provide new levels of information. For instance, in just 3 years GEOSAT provided some 45 million observations of wind speed and wave height and this database constituted the first truly global coverage of these parameters.

Subsequent and continuing observations by the ERS satellites and the Topex-Poseidon mission are building an ever greater source of information for mariners. What is more, as noted in Chapter 2, until the introduction of modern buoy systems, measurements of wave height depended mainly on visual estimates, supplemented here and there by weather ship measurements, and could only provide subjective measures of average conditions. Furthermore, given the high cost of the new automated systems, they have not been able to provide complete coverage and so have obtained correspondingly limited climatological data.

For these reasons we will use satellite measurements where they are available, and note if there are significant differences between these and the standard atlases. To do this, it helps to have some insight into how satellite observations are made and how they can be compared with surface measurements to establish real differences in wind measurements. These discrepancies may not be solely due to the limitations in earlier observations, but could also be due to longer term changes in weather patterns. The subject of climate change is discussed in Chapter 6, but where it appears to be the principal cause of difference between the standard atlases and recent satellite measurements, this will be highlighted in this chapter.

Satellite observations of wave height rely principally on a radar altimeter. This device sends out a sharp very short pulse of microwave energy and then measures the shape of the signal reflected back from the ocean surface. If the surface is flat then the returning signal has a sharp edge and a larger amount of energy is reflected back towards the satellite. If, however, the sea is rough then the reflected radiation travels different distances depending on where precisely it strikes the waves (Fig. 3.1), and the bigger the waves, the more the signal is smeared out. The wave height over the area covered by the radar beam can be calculated from the shape of the returning signal. This calculation has been calibrated against surface observations made by ocean buoys which confirm that radar altimeters are capable of making measurements to an accuracy better than half a metre. These values compare favourably with the accuracy that can be obtained by on-board visual estimates (see Section 5.1.4). Typically satellite observations are combined to produce average monthly values for areas measuring two degrees of latitude and longitude.

Satellite measurements do, however, have limitations. They cannot be made in areas close to land. Furthermore, they do not return to the same spot for some time. In the case of GEOSAT the orbit was designed to return to exactly the same spot every 17 days. So, although there is a lot of overlap between orbits, especially at high latitudes, there are considerable gaps in the coverage which mean it will take a long time to obtain a complete picture of wind and wave statistics. This is a particular problem where for a number of years the conditions are notably different from the longer term average (see next section). Nevertheless, satellite observations provide an unrivalled combination of coverage and accuracy, and are now being used to build up the definitive climatology of current wave conditions. They are of particular value in providing reliable figures for those parts of the oceans which have not been crossed by regular shipping lanes.

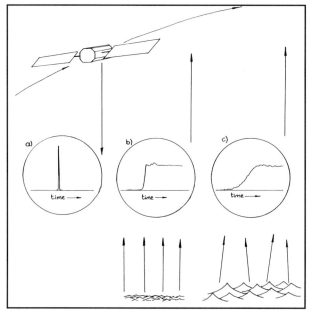

Fig 3.1
A schematic diagram showing how the height of waves can be measured from a satellite by (a) sending out a sharp radar pulse and observing the extent to which the reflected signal from the ocean has (b) a sharp edge when it is reflected by a relatively smooth surface or (c) is blurred out by the ups and downs of a rougher sea-state.

The measurement of wind speeds by satellites is a less precise business. Altimeter data of both wave height and the surface roughness or choppiness, as reflected in the amount of signal returning to the satellite (see Fig. 3.1), can be used to estimate wind speed, the rougher the surface the less signal is returned towards the satellite. In addition other experimental systems can measure both wind speed and direction, but for climatological purposes we have to rely on altimeter data. Although this does not give wind direction, this is not much of a disadvantage in terms of ocean climatology as the broad patterns of the winds are well known.

The real limitation of using altimeters to measure wind speed is the complex process of interpreting the changes in the reflected signal observed by the satellite. This is still the subject of scientific investigation. In the case of GEOSAT, calibration using surface observations suggests that for wind speeds up to 30 knots (15 m/sec) an accuracy of ± 4 knots is achieved, but above 40 knots (20 m/sec) this increases to ± 20 per cent, with a marked tendency to underestimate high wind speeds. Similar results were obtained with comparisons between results obtained from other satellite systems and surface data.5.

The development of tethered and drifting buoys may have attracted less attention than satellites, but in their way they have made equally important contributions to oceanic and atmospheric sciences. Their ability to make continuous accurate measurements of conditions in areas which exert a particularly powerful influence on the weather (e.g. the eastern North Atlantic and the tropical Pacific), is central to this developing knowledge. Perhaps the most important set of buoys is the Tropical Atmosphere Ocean (TAO) array in the equatorial Pacific. This is a key component of the Tropical Ocean Global Atmosphere (TOGA) programme. Completed in December 1994, this array consists of 72 buoys moored every 2° to 3° of latitude between 8°N and 8°S along lines separated by 10° to 15° of longitude. This array has already played a central role in monitoring the development of the most recent ENSO event (see Section 2.7) in the Pacific.

There is just one more refinement to draw upon. This is the extraordinary computing power of modern weather forecasting systems (see Chapter 4) which means all the data being assimilated into these operational activities can be distilled down into up-to-date climatologies. The European Centre for Medium-Range Weather Forecasts (ECMWF) has recently completed a major re-analysis of the data used during the period 1979 to 1993. Where this has been presented in the form of climatological maps which are of interest to mariners, we have used these results in preference to more dated presentations.

In presenting the most interesting new data, the hazards posed by extreme weather will be addressed alongside the broad analysis of climatology. This is best done by first exploring in more detail the conditions in the latitude bands which were introduced in Chapter 2. Then the particular features of certain regions (e.g. the Mediterranean, Baltic, and Great Lakes) will be discussed. Finally, where generic hazards are identified (e.g. tropical cyclones and fog), these will then be presented separately to enable their climatological characteristics to be reviewed more thoroughly.

3.2. Mid-latitudes

Because so much shipping crosses the mid-latitude regions of the North Atlantic and North Pacific, which are subject to marked seasonal patterns, these are an obvious choice for starting our analysis. As noted in Chapter 2, the most important climatological feature of the mid-latitude oceans is the presence of the polar front which defines the general movement of depressions. In the northern hemisphere, the normal pattern is for depressions to form on the eastern seaboard of the major continents and deepen (see Fig. 2.12) as they move north-eastwards.

In the Atlantic they most often form to the south of Newfoundland and run towards Iceland, where they normally achieve their greatest depth before slowly filling as they move into the Norwegian Sea. In winter this process is amplified by two additional factors. First, there is a supply of lows from either the Gulf of Mexico (see Section 3.6) or from out of North America, which pick up new energy off the east coast of the continent. Secondly, the warm waters of the Gulf Stream pump huge amounts of energy into arctic air masses that sometimes stream southwards out across the ocean. This exchange can trigger explosive cyclogenesis with depressions maturing to full-blown storms in no more than 24 hours (see Section 2.5). On the other side of the Atlantic, secondaries form at low latitudes and can run across the UK into Russia, or occasionally dive into the Mediterranean basin.

North Atlantic depressions come in a wide range of intensities and sizes. So it is difficult to categorise their behaviour. Each year between 50 and 100 will follow a variety of tracks across the ocean. They will cover the gamut from minor affairs a few hundred kilometres across and no deeper than 990 mb, to monsters 3000 kilometres across which reach a lowest pressure of below 920 mb. The location of their greatest intensity is equally variable, and can be almost anywhere north of 40°N, but is most likely to be to the south of Iceland between Greenland and the British Isles. On average they reach a lowest pressure of around 980 mb, with lower values in the winter half of the year, and in most years at least one, and often considerably more (see below) will fall below 950 mb during their lifetimes.

Mean wind speeds in the central region (45° to 60°N and 25° to 50°W) in January are 24 to 30 knots, with the highest values being in the south-western part of the region, and the significant wave height (see Section 2.5) for much of the area is in excess of 5 metres (16 feet). These values are somewhat higher than those quoted in the standard maritime atlases and this may be due to the upsurge in wave heights in recent decades (see later in this section). By March these figures have fallen to around 20 to 25 knots and 3.5 metres (11 feet). The area of strongest winds moves north-eastwards towards Iceland in May, when the mean winds are 15 and 20 knots and the significant wave height a little over 2 metres (6 to 7 feet). Throughout the summer the figures remain at a little below these levels, and then pick up rapidly in the autumn, so that by November most of the Atlantic north of 45°N has mean winds in excess of 20 knots and the significant wave height in the central region is over 4 metres (13 feet).

Because of the marked annual cycle, it is difficult to express the frequency of extreme conditions, but as a general indication, north of a line from Cape Hatteras to the English Channel, and including the North Sea and the southern Baltic, will have wind speeds in excess of 30 knots 10 per cent of the time. These stormy conditions will, of course, occur mainly in the winter half of the year and so will be more frequent at this time of year. Significant wave heights in excess of 5 metres (16 feet) will occur 10 per cent of the time in a region 45° to 65°N and 10° to 45°W, again mostly in the winter half of the year.

In the north eastern part of the Atlantic the average figures for winds and waves abate a little. More significantly, the shelter offered by the British Isles reduces the wave height appreciably. So, while the average wind in January to the east of the Shetlands is around 25 knots, and this figure falls to about 20 knots in the southern North Sea, the significant wave height declines from around 4 metres (13 feet) to less than 2 metres (6·5 feet). Again these figures are somewhat higher than given in the US Navy Atlas, and this may be another example of where the satellite data reflects the stormy period of more recent years. The broad features of the shelter offered by the British Isles are, however, evident in the statistics for the rest of the year.

The shelter of the British Isles also creates local patterns of wave heights and wind speeds, with the stormiest conditions being found in a central northern area around 58°N and 2°E and least stormy down between south east England and the Netherlands. Because of the needs of the offshore industry in this region comprehensive statistics of climatic conditions have been published by the Royal Netherlands Meteorological Institute (KMNI). This work is a good example of the regional climatologies produced by national authorities, which offshore industries need to draw on to exploit fully local knowledge when expanding their operations into new provinces.[4]

In the Pacific most depressions form in the China Sea in the neighbourhood of Japan and advance towards the Aleutian Islands, deepening as they go. Normally they then decay in the Gulf of Alaska, but sometimes they pick up steam here and dive south-east towards Vancouver Island and then along the US-Canada border. Secondary depressions can occasionally form north of the Hawaiian islands on the cold front trailing south from a low over the Aleutians. These can then become vigorous systems.

The North Pacific is marginally less stormy than the North Atlantic, but for all intents and purposes the statistics for the region north of 30°N show the same patterns. Such differences are there are relate to the region of strongest winds being a little farther south, running a broad band from around 40°N, 165°E to 50°N, 135°W. The winds in this region are generally about 10 to 20 per cent lower than in the North Atlantic, averaging around 20 to 25 knots in January with a significant wave height of about 4 metres (13 feet), and declining to around 12 knots in the summer with significant wave heights around 1·5 metres (5 feet).

The broad climatology of the movement of North Atlantic and Pacific depressions tells us little about the variations within the seasons, and over longer periods of time. The annual cycle described above is governed principally by the doubling of the strength of the westerly circulation from summer to winter. Together with the southward shift of the polar front, additional storm tracks in winter (see Section 2.4) develop at lower latitudes, notably in the Mediterranean.

As part of this annual cycle, in the eastern North Atlantic around the British Isles the incidence of westerly (defined as lying in the quadrant from south-west to north-west) winds drops from over 50 per cent of the time in January to around 20 per cent of the time in May before rising to around 40 per cent in July and remaining at this level for the rest of the year. At the same time the frequency of anticyclonic weather shows a marked annual cycle with higher values in the summer half of the year, in part as a result of the expansion northwards of the sub-tropical *Azores High*, while the frequency of northerly and easterly winds peaks in the spring. In addition, the incidence of long spells of settled weather shows a clear pattern. The weather is most likely to get stuck into a fixed pattern in high summer (July and August) or mid-autumn (October) and least likely to stay put for weeks on end between April and June, or in September.

As discussed in Sections 2.2 and 2.4, the establishment of fixed circulation patterns are an important feature of the climatology of mid-latitudes of the northern hemisphere. The essential feature is that these patterns exhibit irregular quasicyclic changes which differ markedly from the climatological normals (see Figs. 2.6 and 2.7). These variations can best be visualised in terms of the pressure difference between, say, 35 and 55°N over the North Atlantic. When there is a large pressure difference there are strong sea-level westerlies and a long wavelength pattern in the upper atmosphere wind patterns. With a low pressure difference there is a breakdown of the sea-level westerlies into closed cellular patterns and a corresponding meandering shorter wavelength pattern in the wind aloft.

When examined on a monthly timescale the upper atmosphere patterns at, say, the 500-mb level, at any given time of the year, often do not appear significantly different from the climatological normal. In the northern hemisphere this is partly a consequence of the size of the changes throughout the year, but also reflects the fact that subtle shifts in the patterns can disguise major climatic differences. This fact is illustrated by the examples of the patterns for February 1996 and February 1997 (Figs. 3.2.a and 3.2.b). While the charts have a lot of similarities, the differences which matter are that in 1996 the troughs over eastern North America and eastern Asia are slightly more pronounced, and, more striking, there is a ridge over the North Atlantic and the downstream trough is over the central Mediterranean rather than the Middle East.

The strong westerly pattern in 1997 is effectively a vigorous form of the standard climatological conditions. The westerly flow is particularly strong over the North Atlantic, which experienced an exceptionally stormy month as a series of deep depressions headed north-eastwards. At the same time this strong flow carried mild air deep into Eurasia bringing well above normal temperatures to much of the continent. The more meandering pattern in 1996 brought much colder weather to both eastern North America and northern Europe, with anomalously heavy snowfall extending from Scandinavia down into western Russia.

The patterns at 500 mb are the key to longer term behaviour. These can change suddenly between certain well-defined patterns (*regimes*) which then last from a few days to several weeks. The mix of regimes and their duration defines whether the weather approximates to the climatological normal or gets stuck in a more extreme state. How individual maps of upper atmosphere patterns provide information about the current weather is covered in Section 4.2. In terms of climatological studies what matters is the part played by these upper atmosphere winds in the occurrence of extreme weather events. In this context the meandering patterns tend to be more interesting because they often produce adjacent regions of opposite extremes. The essential feature of

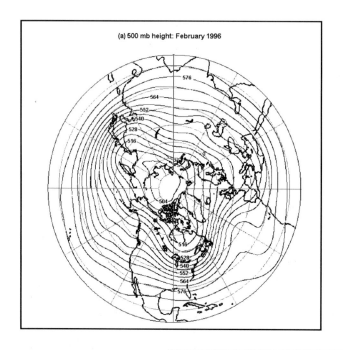

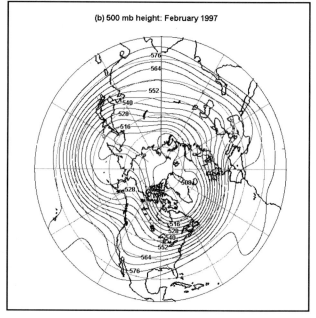

Fig 3.2a & b
Examples of (a) a meandering short-wave pattern at the 500-mb level, which brings marked extremes to different parts of the world, and (b) a stronger westerly long-wave circulation, which brings less extreme conditions.

these conditions is the formation of a *blocking anticyclone* (see Section 2.4) which alters the flow regime of the atmosphere. Over the northern continents blocking brings spells of extreme weather, but to mariners their importance is more to do with the way they alter the storm tracks. A block close to the British Isles diverts low pressure systems northwards into the Arctic or down into the Mediterranean, and often produces unusually quiet conditions in UK waters. Similarly, an eastern Pacific block steers depressions and mild air up into Alaska and southwards towards California. The consequences for ship routeing are substantial, and so, given the prolonged nature of blocking conditions, the ability to predict their onset and continuation is a major challenge to improved weather forecasting (see Section 4.2).

Longer term changes in the incidence of these patterns in recent decades may be a symptom of global warming. The most striking feature has been the upsurge in deep depressions in the North Atlantic in the 1980s and early 1990s (Fig. 3.3). Although this increase may be a temporary development (see Section 6.2), and also partly due to more detailed computer models, whose analysis more accurately represents the true depth of these intense systems in recent years, much of the upsurge is real. Furthermore, more frequent major depressions in the North Atlantic constitute a hazard to shipping and a challenge to offshore operations around the British Isles. In particular, long wavelength swell generated by deep lows is a serious threat to semi-submersible platforms which have a natural heave period of between 20 and 25 seconds. The normal track of these lows means that the long fetch on the southern side, sometimes as much as 3000 km, produces a damaging combination of long period and large amplitude.

Just how extreme these lows can be is seen in the case of the depression in early January 1993 (often known as the *Braer* storm, because it followed immediately after the wrecking of the tanker *Braer* in the Shetlands). This storm, which reached its maximum depth of between 912 and 915 mb during the afternoon of 10 January around 62°N, 15°W (Fig. 3.4) provides a good example of the scale of these depressions. This low had first been identified at 1800 GMT on 8 January, located about 42°N, 61°W, with a central pressure of 1008 mb. At first it only developed slowly, but by 1800 GMT on 9 January it was deepening explosively, with a central pressure of 974 mb at $52\frac{1}{2}$°N, 30°W and had travelled 475 nautical miles (900 km) in the previous 6 hours (a speed of just under 80 knots). Its development was then complicated by coalescing with a second low, to form a single depression which by 0600 GMT on 10 January had a central pressure of 926 mb, having deepened 32 mb in the previous six hours, and was located at $58\frac{1}{2}$°N, 18°W. The observed significant wave

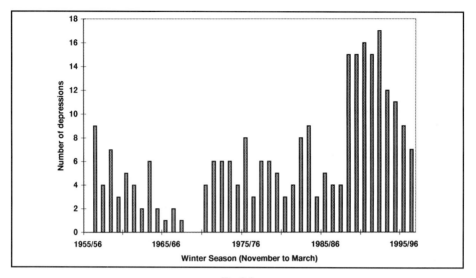

Fig 3.3

The number of very deep depressions (central pressure below 950 mb) for each winter half of the year in the North Atlantic for the period 1960/61 to 1996/7, showing the upsurge of these intense storms in the late 1980s and early 1990s.

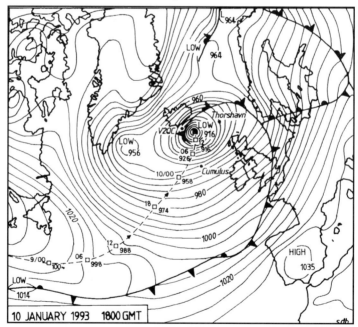

Fig 3.4

The 'Braer' Storm: the Atlantic surface analysis at 1800 GMT on 10 January 1993, showing depression track and central pressure (mb) at six-hour intervals following the identification of the storm as a separate feature 48 hours earlier.   Isobars are drawn every 4 mb (with permission of the Royal Meteorological Society.)

height west of the Shetlands was 17.3 metres (55 feet), with sustained wind speeds of 67 knots. By the afternoon of 10 January the pressure gradient across the North Atlantic was almost 120 mb (Fig. 3.5). Surface winds of gale force or more were reported across the width of the Atlantic from Newfoundland to northern Spain; on the windward coasts of most of western and northern Europe from northern Spain to northern Norway; and in much of the Arctic Ocean between Norway, Iceland and Greenland. Thereafter, the low drifted into the Norwegian Sea and filled over the next three to four days.

Further evidence of the threat such systems pose for offshore operations emerges from statistics obtained during the winter of 1982/83, which was also marked by a series of deep lows. These show that the swell reaching the Norwegian coast can produce sustained very heavy conditions.[6] In particular, periods of up to 12 hours with significant wave heights of the order of 9.1 metres (30 feet) and peak wave periods of 21 seconds (wavelengths of around 700 metres) were observed. A semi-submersible platform at 65°N 8°E registered a heave of 11 metres, indicating the wave periods were close to its natural period. This study also provided interesting results on the complicated wave conditions observed off the coast of Norway. At times there were substantial waves of three different periods (14, 17 and 20 seconds) resulting from locally-generated waves, swell generated within the North Sea and swell coming from the North Atlantic via the 'window' between Scotland and the Faeroe Islands. These extreme conditions were not simply the product of an especially deep low, but the consequence of two lows in close succession. So the climatology of waves in the North Atlantic cannot be expressed simply in terms of the frequency or strength of low pressure systems. To anticipate severe conditions requires detailed forecasts of how the prevailing weather conditions are stirring up the waves.

The average figures disguise the inter-annual variability of the climate. From year to year there can be dramatic changes in the wind and wave regimes, especially in the stormier parts of the world. For instance at Ocean Weather Station Lima, in the North Atlantic (57°N, 20°W) the mean significant wave height during December 1983 and January 1984 was 6·25 metres (20·5 feet), but the following December and January this mean was down to 4·58 m (15 feet): a decrease of 27 per cent between two adjacent years.[7] Changes of this magnitude represent a major shift in the challenges facing maritime operations from year to year, and underline the importance of knowing more about the causes of this variability.

Although events like the *Braer* storm are seen as a consequence of global warming (see Section 6.1) they need to be put into context. While the incidence of exceptionally intense lows did reach unprecedented levels at the end of the 1980s and in the early 1990s, wider analysis provides a more complicated picture. Studies of pressure patterns over the UK region (Fig. 3.5) and the North Sea since the late 19th century show no evidence of any increase in windiness at any time of the year.[8] In contrast, measurements of mean winter significant wave heights in the North Atlantic show an increase of some 80 per cent between the early 1950s and the early 1990s, although changes at other times of the year are much less.[7] So far no satisfactory physical cause has been identified for this rise, in part, because the early data is only available for a few sites. The observed increase does explain, however, why the satellite figures for significant wave heights quoted earlier are somewhat higher than the standard atlases.

Further insight can be gained by examining the interannual fluctuations of pressure patterns in the North Atlantic. A measure of the strength of the westerly winds in this sector is known as the *North Atlantic Oscillation* (NAO). This uses the pressure difference between the Azores and Iceland each winter to record fluctuations in the circulation strength. The values of the NAO over the last 130 years show no appreciable trend (see Fig. 3.6).[9] Instead there are periods of stormy winters followed by quieter runs of years. The most striking feature over the last 100 years being the marked dip in the 1960s and the rise since then. This change may explain the observed increase in wave heights over the last 40 years, but in the absence of earlier statistics, it is difficult to be certain that the more recent rise is simply a reflection of changes in the NAO since the 1960s. This analysis is made more difficult by the fact that the NAO switches back and forth over the years. The return to negative values during much of the winters of 1996 to 1998 may herald another 'flip', and if sustained for a few years will sharply reduce the rising trend in wave heights in the North Atlantic. So understanding what causes the NAO to 'flip' is yet another issue for maritime operations.

The same type of changes occur in the North Pacific, but not necessarily at the same time. Indeed, there is some suggestion in the satellite data that inter-annual variations in the two ocean basins may be out of phase. Years with high waves in the North Pacific tend to coincide with warm ENSO events in the equatorial Pacific (see Section 2.7), while the North Atlantic has low wave heights during these events. An equally striking feature of Pacific pressure patterns is a sudden deepening of the Aleutian Low in 1976 which effectively persisted until 1988.[10] Since then the circulation has returned to the pattern observed before 1976. These changes have been linked also with warming events in the equatorial Pacific and raise important questions for longer term forecasting and climatic change

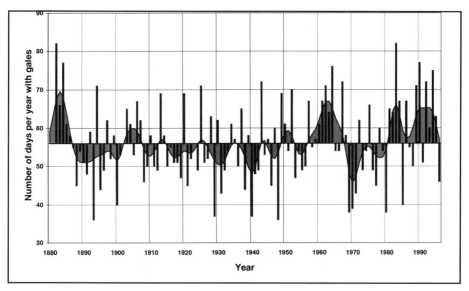

Fig. 3.5.
The number of days with winter gales in the vicinity of the British Isles over the last 100 years, showing no significant increase in recent decades, together with smoothed data showing longer term fluctuations. (Data supplied by the Climatic Research Unit, University of East Anglia, UK.)

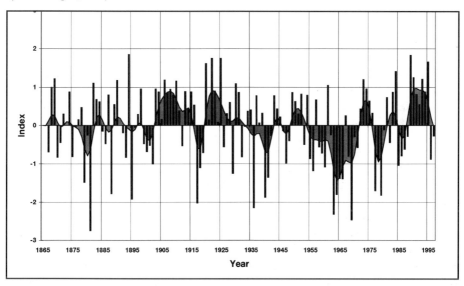

Fig. 3.6.
The North Atlantic Oscillation as measured by the standardised difference of the December to February atmospheric pressure between Punta Delgada, Azores and Stykkisholmur, Iceland, together with a smoothed curve to show fluctuations longer than around 5 years. (Data supplied by the UK Meteorological Office.)

In the southern hemisphere the situation is simpler. The westerlies are much stronger and more constant because of the lack of continental land masses. In winter the mean zonal wind speed between 35° and 60°S is more than twice the peak value in the northern hemisphere around 50°N. Furthermore, while the region of strong winds narrows in summer, the peak value, if anything, increases around 50°S. So the epithets 'Roaring Forties' and 'Screaming Fifties', are fully justified.

The wind and wave statistics for the stormy belt in the southern oceans are on a par with winter conditions over the northern oceans, although they do not match the extremes of mid-winter in the central North Atlantic. So only in July in the southern Indian Ocean around 50°S between South Africa and Australia does the mean significant wave height exceed 5 metres (16 feet) and so match the January conditions for much of the central North Atlantic. These figures are somewhat greater than those quoted in maritime atlases, and this may be a consequence of climatic change, or simply a case of satellites providing improved measurements. It may also reflect the limitations of shipboard observations made in the past in these remote regions.

The winds do not slacken significantly during the summer, with most of the latitude band around 50°S experiencing mean winds in excess of 20 knots throughout the year. The significant wave height does, however, abate a little, falling from values above 4 metres (13 feet) over wide areas, to around 3 metres (10 feet) in January. The only place which experiences any protection from these relentless winds is parts of the South Atlantic in the lee of the Drake Passage and Patagonia. Conversely, on the windward side, the depressions funnelling between southern Chile and Antarctica produce a region of strong but more variable winds.

Perhaps one of the most important consequences for mariners of the waves generated by persistent westerlies in the southern oceans, is the strong swells that propagate to lower latitudes. It is important to be aware that swell follows great circle tracks and is therefore not easy to project over long distances on either navigation charts or on meteorological charts. One particularly dangerous area is where the swell from major storms in the Southern Ocean meets the Agulhas Current along the SE coast of South Africa. Between Durban and East London, which is known as "The Wild Coast", where the continental shelf is at its narrowest, this current is flowing at 4 to 5 knots towards the south-west. If this combination of the current and strong swell from the south west coincides with a low, developing near the South African coast, which then drifts south-eastwards towards the Southern Ocean and deepens substantially, a strengthening and expanding area of south to south-west winds develops on the western side of the low pushing progressively higher and longer swell towards the south-east coast of South Africa.

Even more dangerous is the situation where a marked cold front moves slowly north-east over the coastal waters of south-east South Africa with a strong north-easterly wind ahead of it. These conditions increase the speed of the Agulhas Current, which heightens the probability of abnormal waves. In these circumstances the South African National Research Institute of Oceanology has advised mariners to "stand inshore of the continental shelf edge (where the Agulhas Current will be much weaker) between Richard's Bay and Great Fish Point when steaming towards the south-west with the barometer falling, a fresh north-easterly wind blowing and a change to fresh or strong south-westerly winds forecast within the next 24 hours".[11]

Another important feature of the strong westerlies in the southern hemisphere is that settled spells of weather are much less common than in the northern hemisphere. Blocking does, however, occur, especially in the lee of Australia and New Zealand and to a certain extent downwind of the Andes. These events are less sustained than their northern counterparts, and, given their location, are of little consequence to mariners, apart from around-the-world yachtsmen.

Finally, it is worth noting that, as in the case of the North Atlantic, improved measurements have led to a better understanding of the climatology of these remote areas. After drifting buoys came into routine operation in the Southern Ocean, it was discovered that depressions to the south and south-west of Australia were deeper than had previously been thought. This may, in part, explain why some of the wave height statistics quoted here are higher than the values given in the traditional atlases. But, owing to the shortage of records in the past, it is not possible to make any analysis of whether these differences are in any way related to climatic variations in the Southern Ocean. There are no equivalent series to the NAO or the UK Gale Index for these empty waters.

3.3. Tropics

Apart from being the spawning ground for tropical cyclones (see Section 3.9) much of the weather in the tropics is often regarded as rather dull compared with the dynamics of mid-latitudes. This is because of the small temperature contrasts, the lack of markedly different air masses, and the absence of significant seasonal variations in solar input lead to more subtle effects. Moreover, until recent decades, the huge expanses of the tropical oceans, which occupy so much of these latitudes, had been little studied by meteorologists. But, as discussed in Chapter 2, the tropics are where most of the solar energy is absorbed, and the tropical oceans effectively act as a 'boiler' driving the global weather system. In addition, the discovery that the ENSO plays such a major part in year-to-year fluctuations in the climate has led to a huge upsurge in research in tropical meteorology.

The problem with analysing tropical weather systems, apart from hurricanes, is they involve small surface pressure gradients. These can easily be missed given the relatively large semi-diurnal pressure of about 2 to 3 mb between maxima around 1000 and 2200 hours and minima at 0400 and 1600 hours local time (see Fig. 5.1). The other complication is that close to the equator the *Coriolis force* is small (see Glossary), which means that, unlike at higher latitudes, the wind direction provides no guide to pressure patterns. In many parts of the tropical regions the network of conventional observing sites is limited, adding to the difficulties in making accurate detailed analyses.

Within the broad features of the convergence of the trade winds into the equatorial low pressure region (the *Equatorial Trough*) and the formation of the Intertropical Convergence Zone (ITCZ) (see Section 2.2) there are significant disturbances. These come in various shapes and sizes. The smallest form, with a life span of a few hours, is *individual cumulus* up to a few kilometres in diameter. In fair weather these are generally aligned in 'cloud streets' roughly parallel to the wind direction, or polygonal honeycomb patterns, rather than being distributed at random. In fair weather conditions they have little impact at the surface, but in more disturbed weather they can become intense thunderstorms. These are most likely over warmer waters (temperatures above 26°C [79°F]) and can reach heights of 20 km (65 000 ft) have updrafts of 20 to 30 knots and produce squally conditions at the surface.

On scales up to 100 km cumulus clouds can become grouped into *mesoscale convective areas*, which in turn may cluster together to form a cloud cluster, 100 to 1000 km in diameter. These are of most interest when they develop into more organised systems. Where this clustering behaviour dissipates after only a day or so, the effect at the surface is little different to the impact of individual cumulus.

Where cloud clusters are longer lasting, they can develop into wave disturbances travelling westward in the equatorial and tropical easterly trade winds and are often termed *easterly waves*. Their wavelength is between 2000 and 4000 km and they have a lifespan of one to two weeks, travelling some 6 to 7 degrees of longitude per day. Although easterly waves bear some resemblance to extratropical fronts, they are quite unlike mid-latitude depressions. The weak pressure trough, which constitutes little more than a kink in the isobars, usually slopes eastward with height and the main area of thunderstorms is behind the trough line. So the sequence of weather across a wave is, first, ahead of the trough it is fine with scattered cumulus and some haze. Then, close to the trough line, well-developed cumulus, occasional showers and improving visibility. Finally, behind the trough, the direction of the wind veers in the northern hemisphere, heavy cumulus/cumulonimbus with moderate or heavy showers and a drop in temperature.

Because easterly waves are often a transitional state between disorganised convective activity and the far greater symmetry of tropical cyclones, both their occurrence and definition is open to some interpretation. Furthermore, the movement of the ITCZ during the year means that they are more likely to occur where convection is most active. For instance, in the Caribbean, easterly waves tend to develop in summer and autumn, rather than winter and spring. It is no coincidence that this peak matches the hurricane season (see Section 3.9).

There are two significant localised features of the large scale circulation in the tropics. The first is the effect of the north-easterly winter monsoons of Asia which produce average winds in December and January in parts of the South China Sea of 20 knots, and significant wave heights in excess of two metres (6.5 feet), with a maximum just north-west of Luzon Island, and a secondary maximum south-east of southern Vietnam at about 9°N. In parallel the Australian summer monsoon produces a strong north-westerly flow throughout Indonesia and northern Australia. The absence of lengthy fetches in the Indonesian archipelago means that these winds do not generate substantial waves, although in shallow seas conditions can be sharp and choppy.

The other striking pattern is the strong winds along the southern coast of Arabia from Somalia to Oman from June to September. A product of the south-western Asian monsoon and the huge flow of air up across the Indian subcontinent, these south-westerly winds average over 30 knots. Known as the *Somali jet,* these winds produce the highest mean significant wave heights in the tropics, with values according to satellite measurements in excess of 4 metres (13 feet) during July.

3.4. Polar Regions

In many parts of the world the weather in the oceans that fringe the pack ice and ice sheets of the polar regions is merely an extension of conditions in the wider seas in the mid-latitudes. But there are differences which

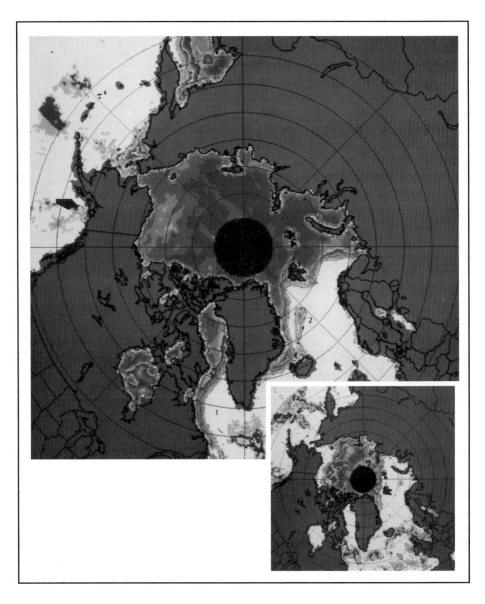

Fig 3.7
A set of satellite images showing the annual expansion and contraction of pack ice in the arctic between late winter (top: April 1979) and late summer (bottom: September 1979) (with permission of NASA).

pose particular challenges to mariners. These relate, in particular, to the annual cycle of the expansion and contraction of pack ice, the dramatic temperature difference between air flowing from the ice fields and the water temperature of the surrounding seas; and to the extreme solar radiation changes from virtually nothing in mid-winter to levels on a par with lower mid-latitudes in mid-summer.

In winter these regions are cold and stormy, with very short days. The Aleutian and Alaskan coastal areas of the North Pacific tend to be north of the main storm track and tend to experience occluded fronts and upper troughs which bring cloudy conditions and frequent snow, but not dramatic shifts in temperature. In the case of Iceland, northern Norway and adjoining areas of Russia, the situation is complicated by frontal depressions moving in to the Barents Sea. So this region, which is navigable to much higher latitudes in winter experiences greater swings in temperature as weather systems move through. This means that not only do temperatures rise to remarkably high levels with the passage of warm sectors of depressions, but the cold plunges that follow can produce severe icing conditions.

The expansion and contraction of pack ice in polar regions is a major factor in governing many aspects of the global climate, but for mariners its impact is of rather restricted interest. In the northern hemisphere, the extent of pack ice varies from a minimum of some nine million square kilometres in late summer, when it is confined to the central part of the Arctic basin, to a maximum of 16 million square kilometres in late spring (Fig. 3.7). At its greatest extent it fills most of the Arctic Ocean, Hudson Bay and the Sea of Okhotsk. It also spills out into the Bering Sea, but it is in the North Atlantic that it has the greatest potential influence. It drifts down the eastern coast of Greenland round Cape Farewell and up towards the Davis Strait where it joins a flow down the coast of Labrador. The annual average extent of the ice in the Arctic is some 13 million square kilometres, of which some 2.5 million square kilometres are open water.

The maximum extent of the pack ice fluctuates from year to year. These variations are closely linked to the strength of the westerly winds during the winter half of the year, especially in the Atlantic sector. When the zonal flow is strong, low pressure systems regularly push through the Norwegian Sea and into the Barents Sea and keep the pack ice back. At the same time this stream of depressions continually pulls arctic air down the Davis Strait and pushes pack ice farther out beyond Newfoundland. During the late 1980s and early 1990s when the North Atlantic was particularly stormy as a result of the positive phase of the NAO in winter

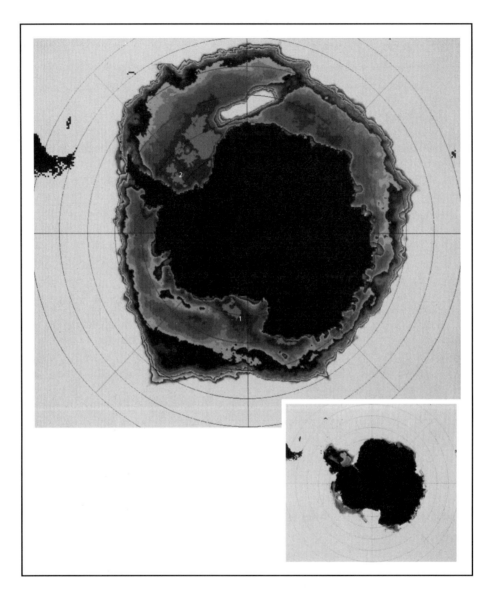

Fig 3.8
Satellite measurements of pack ice provide a continuous record of changes in the extent of the ice in polar regions. This pair of images show the maximum extent of ice around Antarctica in (a) August 1974 and its minimum extent in (b) February 1975 (with permission of NASA).

TABLE 3.1
Classification of icebergs

Descriptive Name	Height (feet)	(metres)	Length (feet)	(metres)
Growler (G)	<17	<5	<50	<15
Small berg (S)	17- 50	5-15	50-200	15- 60
Medium berg (M)	51-150	16-45	201-400	61-122
Large berg (L)	151-240	46-75	471-670	123-213
Very large berg (V)	>240	>75	>670	>230

In addition the icebergs are classified in terms of whether they are non-tabular (NT) or tabular (T). The former include all forms of non-tabular icebergs including dome-shaped, sloping, blocky, and pinnacle. Tabular icebergs are flat-topped with a length-to-height ratio greater than 5:1.

(see Section 3.2), the ice conditions were exceptionally heavy in the West Greenland/Labrador region. This caused substantial problems for shipping in the Gulf of St. Lawrence and along the coast of Labrador during the winters 1991 to 1994. It also produced consistently low sea surface temperatures which were a contributory factor in the disastrous decline in cod stocks in the Canadian Atlantic.

The related issue of icebergs constitutes a much greater hazard to shipping and offshore operations in the north-west Atlantic. It is estimated that 10 to 15 thousand icebergs are calved each year from the glaciers along the coast of west Greenland. A small percentage are carried by the Labrador Current to as far south as 42°N. The service maintained by the US Coast Guard since 1914 (see Section 1.1), shows that the iceberg season extends from March to June. Since 1946 it has averaged 130 days in length. The onset is linked to the start of the break-up of the pack ice at the end of winter. The longest seasons in recent decades were 1972 and 1973, both of which lasted 189 days. Most recently 1991 lasted 183 days - from 23 February to 24 August. The regular classification of iceberg conditions is shown in Table 3.1.

Despite the very real threat posed by icebergs a large concrete oil platform has been successfully installed in the Hibernia Oil Field on the

Grand Banks of Newfoundland. Procedures are in place to tow any threatening icebergs away from the direct line of approach to the platform. Also the platform is designed to withstand strikes from any icebergs which do get through to it.

The amount of navigation in other areas is limited, and the ice is an obstacle rather than a hazard to shipping (see Section 3.7). The Russians have opened up routes in the Arctic Basin along the north coast of Siberia. Proposals to open up the Northwest Passage through the Canadian Archipelago ran into much greater objections in the 1970s. Following the difficult passage of the *Manhattan* in 1969, as a feasibility study by the US petroleum company Humble Oil to see if crude oil could be transported by tanker from Alaska to the east coast of the USA, in April 1970 the Canadian Government brought in legislation to increase the extent of territorial waters and enacted the Arctic Waters Pollution Prevention Act. Although the US government has refused to recognise either of these Acts, it did sign an arctic co-operation agreement with Canada in 1988. So, in practice, the combination of environmental objections, practical difficulties and diplomatic developments means that the North west Passage is unlikely to be opened up as a navigable route in the foreseeable future.

In Antarctic regions the pattern is much simpler. Although the average extent of the sea ice is, if anything, a little less than in the Arctic, being some 12 million square kilometres, it expands and contracts much more. This gigantic annual pulse, from a maximum extent in August being nearly 20 million square kilometres to a minimum of only 3 to 4 million square kilometres in February can be seen clearly in Figure 3.8. As with the Arctic, within this ice boundary there is some 3 million square kilometres of open water in the form of leads and, occasionally, much larger areas of open water known as 'polynyas' (an example can be seen in Fig. 3.8a in the eastern part of the Weddell Sea).Although the expansion and contraction of sea ice has major consequences for the climate of the southern hemisphere, it does not constitute a significant problem for normal shipping as so few vessels venture into the region, apart from those supplying bases in Antarctica. The same detached view can be taken of the huge tabular icebergs which are calved from the ice shelves around the continent, notably the Ronne-Filchner Ice Shelf in the Weddell Sea. These can reach sizes of 90 kilometres by 90 kilometres and stand nearly 50 metres (165 feet) out of the water. They may survive for several years slowly breaking up as they move to lower latitudes. In an extreme instance in 1894 an iceberg was recorded as far north as 26°S, but for the most part they only represent a remote threat to unwary shipping in the more inhospitable regions of the southern oceans. If, however, the continental shelf around the Falklands is developed as an oil province then offshore structures will have to be designed to accommodate this potential risk. The experience gained in operating the Hibernia Platform on the Grand Banks of Newfoundland should prove to be invaluable.

The other major hazard in polar regions is ice accretion. This can occur wherever the combination of low air temperatures and high winds occur in navigable waters. These are found most frequently close to the edge of pack ice, or where exceptionally cold air streams off ice or snow-covered land masses. The most dangerous conditions occur when the air temperature is low enough for freezing spray to encrust and dangerously overload the superstructure of small vessels (100 to 1000 tons). There is a long history of losses of sailing ships around Cape Horn, and whaling boats and seal catchers in arctic waters. The most frequent examples of losses in recent decades are fishing vessels in the seas around northern Japan. Other dangerous areas are the Bering Sea, around Newfoundland, the Barents Sea and north of Iceland. The last named area saw the loss of two large British trawlers (*Rodbrigo* and *Lorella*) in January 1955, and four more in January and February 1968 (*St. Romanus, Ross Cleveland, Kingston Peridot* and *Notts County*). The latter disaster led to an official Commission of Enquiry[12] and also led to the publication of a WMO Report[13] on ice accretion. What follows draws on these sources.

The rate of ice accretion due to freezing spray depends on the air temperature, wind speed, and sea surface temperature. The colder the air, the stronger the wind, and the colder the sea, the more severe the icing (see Fig. 3.9). As can be seen, icing only occurs when the air temperature is below -2°C (28.4°F), and is not a major problem for wind speeds below Force 7, and when the water temperature is above 8°C (46.4°F), but the risks rise rapidly as the conditions deteriorate. So, when temperatures well below -2°C (28.4°F) and strong winds are forecast then it is wise to avoid cold exposed waters. For instance, the *Arctic Pilot*[14] noted that in the Denmark Strait north-easterly gales lasting for several days with temperatures below 29°F and winds not uncommonly Force 12, can produce high seas in excess of 50 feet with rapid accretion of ice, and advises "seek warmer water and the shelter of the West Coast of Iceland".

As far as specific weather features are concerned, perhaps the most interesting local phenomena are *polar lows*. These small, intense depressions occur where bitterly cold air streams off polar ice or the ice sheets of Greenland or Antarctica out over relatively warm water. Normally these airstreams form lines of intense convection (often termed cloud streets - see Fig. 3.18), but when the conditions in the upper atmosphere exhibit a tendency to form a cold pool then this activity can

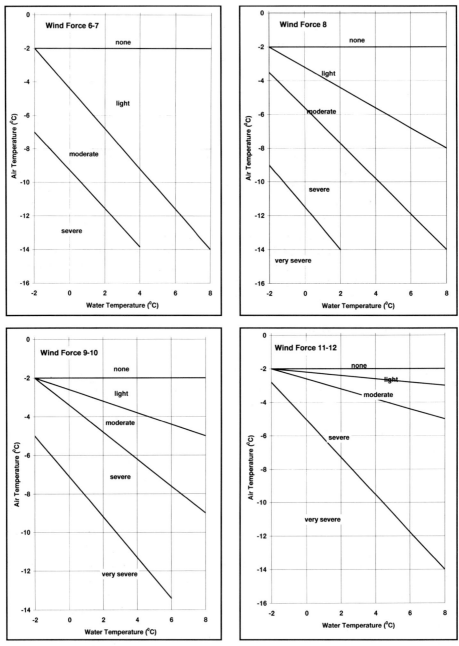

Fig 3.9
A set of diagrams showing how the rate of ice accretion on ships' superstructure depends on wind speed, air temperature and water temperature. The ice build-up rates are: i) light = 1-3 cm/24 hours, ii) moderate = 4-6 cm/24 hours, iii) severe = 7-14 cm/24 hours, and iv) very severe ≥ 15 cm/ 24 hours.

develop more organised circulation and coalesce into extraordinarily severe local storms. Typically these systems will be 400 to 800 km across, but in some instances they may be no more than 100 km in diameter (see Fig 5.3). In spite of their small scale they can produce wind speeds up to 60 knots, and when combined with either heavy continuous snowfall or showers, they represent a severe hazard to shipping.

Polar lows can occur wherever there is frigid arctic air flowing out over adjacent oceans. They are common in the Labrador, Greenland and Norwegian Seas, being most frequent where the air has an unimpeded flow off the ice, either down the Davis Strait or out between Greenland and Svarlbad towards Norway. They are relatively less frequent in the Bering Sea but occur more often to the east of Sakhalin and Kamchatka where Siberian air flows out to sea. They are also a feature of the southern oceans around Antarctica. While these systems are primarily found close to the edge of the polar pack ice, they can also occur as secondary features of large mature depressions which drift to high latitudes. In this form they can be steered to lower latitudes, notably down in the North Sea and the Baltic (see Section 5.4).

3.5. The Mediterranean

Although the Mediterranean gives its name to a broad climatic domain, which features mild wet winters and hot dry summers, the Mediterranean basin itself has a rather more complicated climatology. This is because, unlike other parts of the world which have a similar climate (e.g. California, central Chile, southernmost tip of South Africa and southern Western Australia), the Mediterranean sea is surrounded by land masses. This does not alter the archetypal hot dry summer and mild wet winter. But, in winter it produces a far more dynamic regional climate, which poses major challenges for shipping. So it deserves a section of its own.

In ancient times, seafaring was restricted to the period of May to October. More bluntly, in the Middle Ages the winter half of the year was the time to prepare for war, which was waged in the summer. The switch between summer and winter usually comes quite suddenly in late October as the Azores high declines and its influence over the Mediterranean effectively ceases.[15] By this time the Eurasian land mass to the north is cooling rapidly while water temperatures across the Mediterranean basin remain high. This means that cold plunges of arctic or polar air which reach the basin are much colder than these waters and this is the recipe for vigorous convection. These conditions happen most often when maritime polar air is pulled southwards behind a combination of low pressure over Scandinavia and also over northern France and western Germany (Fig. 3.10).

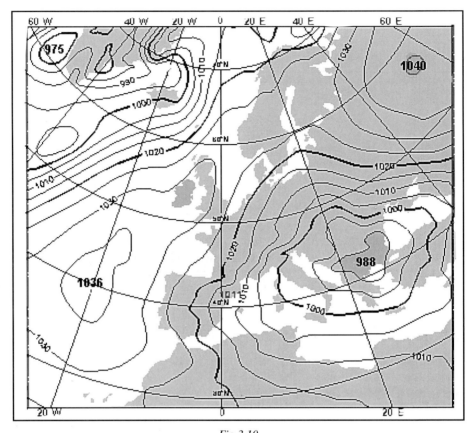

Fig 3.10
The synoptic conditions which often lead to the Mistral blowing down the Rhone valley and out into the Gulf of Lions. This was the situation forecast for 14 February 1996 by the ECMWF on 9 February 1996 (with permssion of the ECMWF.)

There is, however, a much more significant aspect of incursions of cold air into the Mediterranean basin: this is cyclogenesis. The combination of warm water and the mountainous terrain on the northern edge of the basin, provide the ideal conditions for spawning low pressure systems, notably in the Gulf of Genoa, to the south of the Ionian Sea, and around Cyprus (Fig. 3.11). This means the Mediterranean is exceptionally stormy in the winter. Furthermore, because almost all of these storms are generated locally (less than 10 per cent of the low pressure systems enter from the Atlantic) their development is often rapid and this requires forecasts which accurately reflect the regional conditions. It also places a high premium on local knowledge in interpreting forecasts.

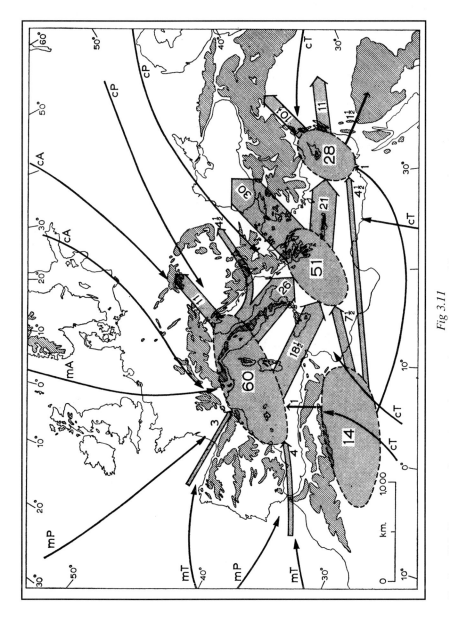

Fig 3.11

Tracks of Mediterranean depressions, showing the average annual frequencies, together with air mass sources (after 'Weather in the Mediterranean' HMSO 1962; Crown Copyright Reserved).

Fig 3.12
Meteosat image of a storm in Mediterranean on 16 January 1995 showing the remarkably symmetrical circulation similar to a tropical cyclone. (© 1999 EUMETSAT)

Occasionally these winter storms can behave remarkably like a tropical cyclone (see Section 3.9), although the fact that they occur in winter over relatively cool water (around 15°C [59°F]) has led to suggestions that they might be more akin to polar lows (see Section 3.4). Whatever their designation, they represent a serious threat to shipping. A good example occurred in 1995, when a low pressure system formed near 38°N and 14°E on 13 January and moved east and then south over the next two days, reaching a minimum pressure of 990.6 mb during 14 January. By midday on 16 January, from space, it looked just like a tropical cyclone (Fig 3.12), with a distinct 'eye', and the peak winds were estimated at 75-80 knots earlier that day. A similar event on 26 January 1998 hit a container ship at 37°N and 18°E with winds up to Force 11 and mountainous seas. She lost 37 containers which subsequently proved to be a serious hazard to navigation.

The pattern of cyclogenesis is reflected in the wind and wave climatologies. The peak in mean winds of nearly 25 knots occurs in January to the east of Corsica and Sardinia, and, when depressions develop rapidly, can produce hurricane-force winds between the two islands. A secondary area of strong winds at this time of year is found in the area from the southern Aegean Sea to south of Crete, where mean winds of 25 knots are experienced. The significant wave heights are, however, modest because of the shelter provided by the surrounding land masses. So mean values are around 1·5 metres (5 feet). The area of strongest winds moves into the Gulf of Lions by March and declines to around 20 knots. It then drifts towards the coast of Spain by May declining all the while. In summer the strongest winds are associated with more local effects (see below). In autumn the strongest mean winds (up to 25 knots by November) occur in the Straits of Gibraltar, but by then much of the open Mediterranean is experiencing mean winds of around 15 knots.

A secondary feature in understanding the climate of the region is the role of local winds. Some of these, like the *Mistral*, are part and parcel of larger scale weather systems already discussed, others are more localised. The Mistral, which blows down the Rhone Valley and out in to the Gulf of Lions, is usually associated with the weather pattern presented in Fig. 3.10. In winter this wind is reinforced by regional cooling effects of snow-covered upland areas which produce a downward flow of air from the Alps and the Massif Central. This gravitational sinking of cold dense air down into the valley bottoms is known as *katabatic wind*. This cold dry air draining down into the Rhone Valley, when combined with a stronger northerly synoptic pattern can produce wind speeds of 80 to 85 knots in the vicinity of the Rhone delta with bright, clear conditions. Between December and May there are on average about 26 days when the Mistral blows with wind speeds of 33 knots or greater, with a slight peak in activity in March and April (11 days between them). For the rest of the year there are only rare episodes although less intense events occur throughout the summer. These winds usually decrease rapidly seawards but may occasionally extend to Malta and North Africa. Their danger to shipping is, however, to raise a heavy sea in a short time. This is particularly true in March when the average significant wave height in the southern Gulf of Lions rises to 2 metres (6·5 feet), the highest value experienced in the Mediterranean.

Similar katabatic effects are associated with the *Bora* which is a north-easterly wind that occurs along the eastern shore of the northern Adriatic in winter where it produces violent squalls and gusts sometimes reaching 100 knots. These conditions are most vigorous when the Mediterranean low is well developed and persistent high pressure over Europe. In particular, behind a cold front moving south-east over the Adriatic the effect is reinforced by the katabatic descent of cold air from the mountains of Dalmatia. But, in spite of these conditions being reasonably well defined, it is the suddenness with which they develop that constitutes the greatest danger to shipping.

TABLE 3.2.
THE WINDS OF THE MEDITERRANEAN

Name	Description
Bora	A cold, very dry, north-easterly wind, often violently gusty and squally, which blows down from the mountains on the eastern shore of the Adriatic. Most common in winter.
Ghibli	A hot, dry, southerly wind which blows over Libya ahead of depressions moving eastwards in the Mediterranean. Most common in late spring.
Gregale	A strong north-easterly wind in the west and central Mediterranean in the winter half of the year: particularly important to Malta with its north-facing harbours.
Khamsin	A hot, dry, southerly wind which blows over Egypt ahead of depressions moving eastwards in the Mediterranean. Most common in late spring.
Levante	An easterly wind in the Straits of Gibraltar. Most frequent from July to October and in March.
Leveche	A hot, dry southerly wind which blows onto the south-east coast of Spain ahead of an advancing depression.
Llevantades	Potentially dangerous north-easterly gales which blow along the east coast of Spain in the Spring and Autumn associated with slow moving depressions crossing the western Mediterranean.
Meltemi (Etesians)	Northerly winds which blow during the summer in the the Aegean.
Mistral	A northerly wind which blows off shore along the north coast of the Mediterranean from the Ebro to Genoa. Can occur at any time of the year, but strongest and most frequent between December and April.
Sirocco	Warm, dust-laden southerly wind from the Sahara, most common in spring and autumn, usually occurs ahead of a cold front.
Vendavales	Strong squall south-westerly wind in the Straits of Gibraltar and east coast of Spain. Associated with depressions and so a feature of the winter half of the year.

Along the east coast of Spain the *Llevantades* and *Vendavales* are part of the shift in peak wind patterns during the winter half of the year. The first features gales from north-north-east and east-north-east usually associated with depressions crossing the Mediterranean between France and Algeria, and can produce particularly heavy seas when the wind is in the north-east and has a long fetch. These gales are most frequent and dangerous in autumn and, especially in spring. The Vendavales features strong south-westerly winds through the Straits of Gibraltar and up the coast of Spain associated with advancing depressions from late autumn to early spring. Their principal danger is violent squalls and thunderstorms. Other examples of local winter winds and their characteristics are listed in Table 3.2.

The transition between the two halves of the year, as described earlier, completely alters the character of the local winds. In the absence of cyclogenesis the patterns of local winds is dominated by where the air is coming from and what happens to it along the way. For instance, the *Sirocco*, (known as the *Leveche* in Spain, the *Ghibli* in Libya, and the *Khamsin* in Egypt), the hot southerly wind associated with the advance of a depression moving eastwards across the Mediterranean, is most common in spring as the subtropical high moves northwards. Since it draws hot, dry, dust-laden air from the Sahara, on crossing the sea it picks up copious moisture producing a disagreeable combination of heat and humidity. The most significant consequence of the Khamsin is that at any one place it often sets in suddenly, with little or no advance signs, rather than building up gradually. This is a serious hazard for some marine operations, such as mooring tankers at offshore loading terminals. These problems are compounded by the fact that the Khamsin is often a relatively narrow stream of wind. Because of this it is not easy to deal with in broadscale marine weather forecasts which are unable to provide the necessary local detail. The other concern for shipping is that as it passes over the cooler waters of the northern Mediterranean in spring and early autumn it readily forms thick fog.

Perhaps the best known summer winds are the *Etesians* (the *Meltemi* in Turkey) which blow from between north-west and north-east in the Aegean Sea. These are a consequence of the low pressure system that stretches from Anatolia to north-west India which is formed by the intense heating of the region (a 'heat-low'). These peak in August when the mean wind speed is round 15 knots over the southern Aegean. Although they produce choppy conditions for much of the time, their principal benefit is to moderate temperatures in coastal regions. Occasionally, they may, however be associated with violent thunderstorms, which produce sudden wind and squalls (known as

Bourini in Greece) and cause considerable damage to local shipping. These are most likely to occur when the circulation pattern combines with an influx of cold air at upper levels from more northerly latitudes while at the surface the hot air continues to be drawn round the heat low. This combination makes the atmosphere highly unstable and hence ideal for convective activity and the formation of vigorous thunderstorms.

3.6. Gulf of Mexico and the Caribbean

As the principal source of maritime tropical (mT) air in eastern North America the Gulf of Mexico plays a major role in the climate of this side of the continent. But for mariners the region is better known for three specific reasons. These are:

i) it is an important area for hurricanes from around late June to early November (this subject is dealt with in Section 3.9);

ii) it is an area of vigorous thunderstorm activity, especially around Florida in the summer; and

iii) it is an area of vigorous cyclogenesis in the winter half of the year.

The reason for the occurrence of vigorous cyclogenesis is both the contrasts in temperature between the air masses that may collide along the polar front that sometimes extends back into the Gulf of Mexico during the winter, and also the strong gradient in sea surface temperatures between the cold continental shore and the tropical waters of the Caribbean. The temperature gradient is particularly marked off the Mississippi delta in late winter when cold water flowing into the Gulf can produce temperature differences as great as 14°C (8°C to 22°C) over only 20 kilometres. This combination can produce explosive results. Perhaps the best known example of this in recent years was the "Storm of the Century" in March 1993 (see Fig. 3.13). This storm is not only a good example of how quickly these storms can develop, but also how they often continue up the East Coast (see Section 3.2) posing a threat to shipping in both areas.

The details of the storm show just how dangerous this rapid deepening of low pressure systems in the Gulf of Mexico can be. In the 24 hours from 00.00 GMT on 13 March the central pressure dropped from 989 mb to 960 mb as the storm moved from the western Gulf to the Chesapeake Bay. This qualifies the storm as a *meteorological bomb* (see Section 2.5). As it moved from Florida it was preceded by a severe squall line which generated hurricane force winds (highest gusts being 96 knots), severe thunderstorms and tornadoes, and a storm surge of up to 3 metres

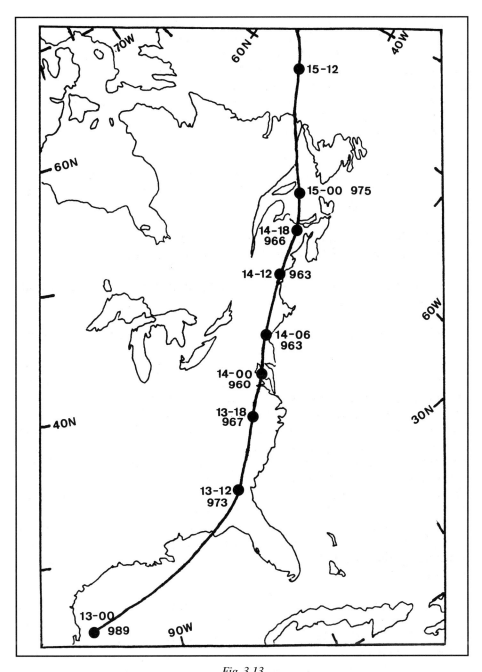

Fig. 3.13
The track of the "Storm of the Century", March 12-14, 1993. The figures indicate the time (date-time GMT (EST + 5 hours)) and central pressure (mb) of the depression (with permission of the Royal Meteorological Society.).

(10 feet) in the Apalachicola area. The US Coast Guard rescued 235 people from 103 vessels in the Gulf of Mexico.

As the storm moved up the East Coast it broke a number of all-time record low pressure figures, most of which had been established by hurricanes. The lowest value recorded was 958·4 mb at White Plains, New York. While onshore the most significant feature was the massive snowfall, at sea it was the winds. These can be gauged by the fact that at Kennedy Airport in New York City on the afternoon of 13 March there were sustained winds of 55 knots, with gusts to 67 knots, while on Cape Breton Island, Nova Scotia, a peak gust of 114 knots was recorded early on 14 March.

The wave heights were just as dramatic.[16] Buoy number 41002, SSE of Cape Hatteras, measured a significant wave height of 15·7 metres (51 feet). Available climatological data has suggested that the 100 year extreme for the area was 12 metres (40 feet). Farther north, off the coast of Nova Scotia, 160 kilometres (100 miles) south of Sable Island, on the morning of 15 March, the 600 foot Liberian registered gypsum carrier *Gold Bond Conveyor* sank with the loss of all 33 crew, having radioed that it was taking on water in 20 metre (65 feet) waves. This figure was confirmed by Buoy number 44317, some 320 kilometres (200 miles) southeast of Nova Scotia which measured significant wave heights greater than 16 metres (52 feet) and maximum wave heights of greater than 30 metres (around 100 feet). These figures were reckoned to be 50 per cent higher than the calculated 100 year extremes for the area.[16] Subsequently the Topex/Poseidon satellite measured average waves of 12 metres (40 feet) travelling across the mid-Atlantic from the storm, confirming that the swell from such storms poses a threat far and wide.

These figures might be seen as confirmation of the claim of the March 1993 event to be the 'Storm of the Century' but for the fact that similar wave heights had been recorded by Buoy 44317 only 17 months earlier during the 'Halloween Storm' of 1991. This storm was equally interesting for a number of reasons. It was formed from the interaction of a late-season hurricane, *Grace*, and an extratropical depression. Although this combination remained well off shore of the Atlantic coast of the USA, between Capes Cod and Hatteras, it stooged around for over 48 hours on 29 to 31 October 1991. The erratic behaviour combined with the highly unusual feature, that for much of the time the storm was slowly travelling from east to west, meant that the forecasts of its intensity and position were not good. The sustained wind fields of this quasi-stationary system generated the massive waves observed by Buoy 44317, and a storm surge which did immense damage to coastal communities from Maine to

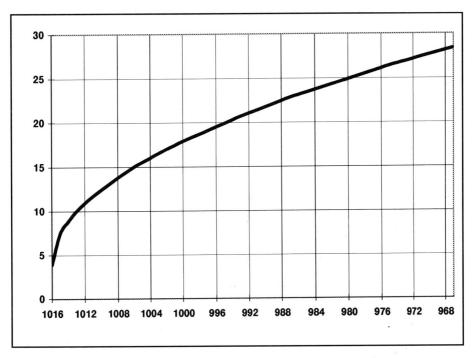

Fig 3.14
Relationship between the central pressure and wind speed of winter storms in the Gulf of Mexico.

as far south as the US Virgin Islands. The insured losses in the US were estimated at between $1·5 and 2 billion. What these two storms confirm is that mariners need to recognise that during major storms very extreme sea states may be much more common than might be expected on the basis of traditional statistics (see Section 2.5). The reasons for this are explored in more detail in Section 4.5.

As a measure of the threat of winter storms in the Gulf of Mexico an empirical relationship has been derived from the analysis of 26 storms.[17] This is shown in Fig. 3.14. The important feature is that developing storms rapidly develop significant windspeeds, and although these grow less rapidly with the deepening of the depression, they represent a major threat to shipping.

3.7. The Baltic

In many respects the Baltic is a sheltered version of the North Sea and hence an extension of the North Atlantic when it comes to wind and wave statistics. There are, however, two features worthy of separate consideration. First, the Scandinavian mountains influence the passage of fronts associated with Atlantic depressions moving eastwards to the north of the Baltic. In effect, the mountains delay the progress of the fronts and, in certain circumstances, this can lead to cyclogenesis in the vicinity of the Skaggerak, especially in the winter half of the year. These rapidly developing small, but intense, lows are a potential hazard and so forecasts of rapidly developing stormy conditions in the southern Baltic have to be taken seriously.

The second feature is the formation of ice in the winter. The season ranges from four to seven months and is highly variable from year to year. In mild winters, as were common in the late 1980s and early 1990s, any icing will be restricted to the north of the Gulf of Bothnia. In cold winters, like 1985 and 1996, much of the Baltic will have heavy ice which seriously restricts shipping.

The average figures are that the northern half of the Gulf of Bothnia, and the eastern end of the Gulf of Lagoda, is frozen during some part of 90 per cent of winters. The southern part of the Baltic down to around Stockholm has some cover in half of the winters while south and west of Gotland out to the Skaggerak has ice cover in only about 10 per cent of winters with the deepest water south of Gotland being the most likely to remain ice-free. Between 1980 and 1993 the maximum ice cover ranged from 12 per cent (52 000 km^2) to 96 per cent (405 000 km^2) with a mean cover of 181 000 km^2 and in the coldest winters like 1966 and 1970 the cover can be total.

3.8. Great Lakes

In many respects the Great Lakes represent a rather exaggerated microcosm of the issues described for the Mediterranean and the Baltic. Again the effect of being land-locked dominates the climatology of the region, but being situated on the eastern side of North America the climate is far more continental. This means hotter summers than the Mediterranean, and winters every bit as cold as the Baltic, in spite of being some 15 degrees of latitude farther south. When combined with the fact that eastern North America is a meteorological battleground as either frigid air can sweep down from the Arctic or warm humid air can be drawn up from the Gulf of Mexico. These factors are the reason why two weather phenomena are of real concern to mariners crossing the Great Lakes: November/December storms and winter ice.

During the autumn, as the land mass cools, the Great Lakes stay relatively warm. For the inhabitants of the southern shores of the lakes this means heavy local snowfall when arctic winds scoop up copious moisture from the warm waters and dump it as snow on the leeward land. But for mariners the greatest threat is autumnal depressions spawned in the Gulf of Mexico, which run up toward Labrador picking up extra energy from the Great Lakes on the way. This effect is greatest when the lakes are relatively at their warmest compared with the surrounding land which explains why many of the greatest shipping disasters have been in November. In 1975 on 10 November the large bulk ore carrier, *Edmund Fitzgerald*, loaded with 26,000 tons of taconite pellets, sank suddenly with all hands during an intense storm. In November 1913, on Lake Huron alone, eight vessels foundered during a great storm with the loss of all 250 crew. The interesting feature of these severe early winter storms (central pressure $\leq$ 992 mb) is that they have almost doubled in frequency during the 20th century. So they seem to represent an increasing threat to shipping.

By late December the temperature of the lakes has normally dropped appreciably and ice becomes the challenge for mariners. Lake Erie, being the shallowest usually freezes over completely, whereas Lake Ontario normally has little ice cover and much of Lake Michigan remains ice free. By comparison Lake Superior is largely frozen over except for around the eastern end of the lake, while Huron is a little less fully covered. The partial nature of this ice cover means that maintaining shipping depends on not only the severity of the winter but improving technology to detect gaps in the ice and to keep shipping lanes open. The increased use of both satellite images and ground-based synthetic aperture radar has improved the detection of possible routes and altered operations. Prior to satellites the Lakes used to be shut on average for about two months a year. Now, even in severe winters like 1978 and 1994 they are not totally closed. So, in terms of disruption of shipping it is not possible to make any observations about whether there have been any significant climatic changes in the severity of ice conditions in recent decades.

3.9. Tropical Cyclones and Hurricanes

Tropical cyclones and hurricanes, in their various forms (see Section 2.2), occur across all the tropical oceans, except the South Atlantic (see Fig. 2.18).[18] They have a well defined season which is linked to when sea surface temperatures (SSTs) are at their highest levels. So it is best to review their climatology as a whole, before considering how their behaviour varies from place to place.

First, we need to expand on the general observations made in Chapter 2, and compare tropical cyclones with extra-tropical depressions which have been discussed earlier in this chapter. Essentially, at the surface, they are both low pressure systems having identical circulation properties. The two important differences are that well-defined tropical cyclones are more symmetric, lacking the frontal systems of their mid-latitude cousins, and experience stronger winds. The latter is principally a matter of pressure gradient - a typical extra-tropical depression with a central pressure of 950 mb may have a diameter between 2000 and 3000 kilometres, whereas a tropical cyclone of similar depth is often less than half this diameter.

As both Tables 2.3, and 2.4 show the central pressure is a measure of the maximum windspeed, and indeed empirical relationships have been constructed to estimate wind speed in terms of central pressure of intense hurricanes (Fig. 3.15). But as can be seen from the scatter between the figures in these various sets of observations, this is by no means a precise measure. So this relationship must be used with caution, as the Admiralty Weather Manual[19] notes "the barometric minimum is not necessarily a good criterion for storm intensity." Small, rapidly developing tropical cyclones can spring surprises. It notes a storm struck Guam in the Pacific Ocean in March 1923 with winds up to 134 knots only produced a

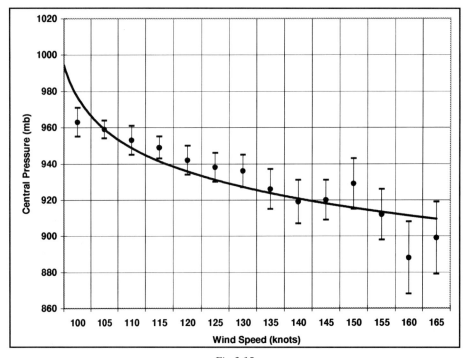

Fig 3.15
The relationship between central pressure of intense Atlantic hurricanes and the wind speed during the period 1970 to 1989

minimum barometric reading of 987 mb. While the centre of the storm did not pass directly over the recording station, the important point is "it is not necessary for the bottom to fall out of the barometer before winds of hurricane force may be expected." Within intense tropical cyclones, in spite of their symmetry, there may be a limited core of extremely high winds. Hurricane *Andrew* is an excellent example of an intense storm which caused immense damage only along a narrow corridor of around 10 kilometres wide. Less intense storms, while not exhibiting such limited ferocity, can affect a much wider area. But, as a general rule, if the glass heads below 980 mb in the tropics - then watch out.

In defining the climatology of hurricanes, it is best to start with the Atlantic basin (the tropical North Atlantic, the Caribbean and Gulf of Mexico). Although this is not the region where most tropical storms occur, it is where the most lengthy and detailed studies have been conducted.[20] Data sets extend back to 1886 with reliable data of all hurricanes to affect the United States since 1899. More important, in 1944 the US Air Force and Navy began the practice of routine reconnaissance flights into the storms. Prior to this, unless a tropical cyclone went directly over a ship or coastal station, the exact intensity of the storm was unknown. Since then, storms that have threatened coastal areas have been kept under nearly continuous aircraft surveillance. Furthermore, from the mid-1960s continuous satellite coverage has ensured that every tropical cyclone in the Atlantic basin has been monitored. So while before 1944 a number of short-lived tropical cyclones in the eastern and central Atlantic were completely missed, and a few may also have gone undetected until 1967, the coverage over the last century is remarkably high.

The hurricane season in the Atlantic basin extends from 1 June to 30 November. Activity outside this period is negligible. Both named storms and hurricanes show a strong maximum in mid-September, with most cyclones occurring between 1 August and 31 October. This late summer/early autumn maximum coincides to the time of the largest areal extent of high SSTs and the lowest wind shear in the troposphere over the tropical Atlantic (see Section 4.7). Intense hurricanes with maximum sustained windspeeds in excess of 100 knots (category 3, 4 or 5 on the Saffir-Simpson scale - see Table 2.3) have a much sharper peak (Fig. 3.16) with 57 per cent of all such storms occurring in September. The average number of intense hurricanes each year since 1944 has been 2·2, with the number of days (defined as the sum of six-hour periods when intense hurricanes were in existence) experiencing these conditions being 4·6. The figures do, however, fluctuate dramatically from year to year, ranging from none in quiet years to as many as eight intense hurricanes and nearly

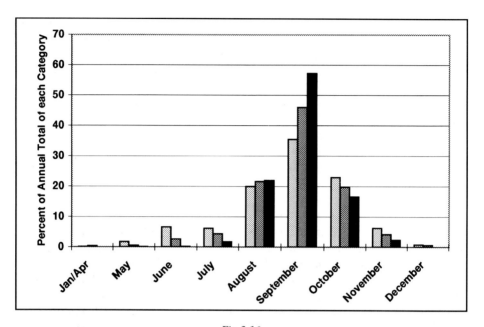

Fig 3.16
The distribution of intense Atlantic hurricanes (on the Saffir-Simpson scale, where category 3 is shown as light bars, category 4 as medium bars, and category 5 as dark bars) during the season.

25 days of winds above 100 knots somewhere in the Atlantic basin in the stormiest seasons.

The tracks of intense hurricanes follow the broad pattern shown in Fig. 2.17, but show interesting variations during the season. From mid-June to late July they are infrequent, and when they do occur, they tend to form in the Gulf of Mexico or the western Caribbean Sea and generally track towards the north or west. From August to early October, the intensification of cyclones into intense hurricanes can occur nearly anywhere in the central and western Atlantic basin between 10° and 35°N. By mid-October, however, their formation is restricted almost exclusively to the western Caribbean or the Atlantic between the Bahamas and Bermuda, and typically begin taking a due north or north-east course, though occasionally moving into the East Coast of the United States.

Because of changing levels of measurement it is difficult to draw hard and fast conclusions about how the frequency of intense hurricanes in the Atlantic basin has changed over the last century. But, after careful analysis of the records, it is clear that on any basis, including number of intense hurricanes, the time with sustained winds above 100 knots, or the maximum sustained winds, activity was much higher in the 1940s to 1960s than it has been since (see Fig. 1.2). But, 1995, 1996 and 1998 were all active seasons, and if sustained, this upsurge may reverse the general pattern of decline since the late 1960s. Nevertheless between 1970 and 1991 the average number each year fell some 40 per cent (from 2·7 to 1·6), as compared with the period 1944 to 1969, and the number of days each year dropping by more than half (from 6·2 to 2·7). The implications of this decline for both the seasonal forecasting of hurricanes and the impact of global warming will be discussed in Chapters 4 and 6. Turning to other parts of the world, the state of knowledge is less complete, although there is quite a lot of information to draw upon.[21]

The annual number of tropical cyclones in the northern Pacific is greater than in the Atlantic basin because there is a far larger area of SSTs in excess of 27°C in the warmest months. This more extensive energy source means the season extends into May and December and no month is completely free of tropical storms, as *Hannah* in January 1997 demonstrated. Because of the distribution of SSTs, the ocean is usually regarded as being divided into two areas - the eastern North Pacific (east of around 160°W) and the western North Pacific (west of 160°W). The eastern area can be very active, with an average of nine tropical storms achieving hurricane status each year. Most of these never make landfall, but cross the North Pacific in the latitude band 10° to 25°N before petering out. Occasionally they recurve back into the central American mainland or even the south-west United States. Even more rarely they reach the Hawaiian islands, as in the case of the devastating hurricane *Iniki* in September 1992 which caused well over $1 billion in damage. This category 4 hurricane was the strongest tropical cyclone to hit the islands in at least 90 years. The islands are usually protected from significant tropical cyclones by the cooler waters that surround them.

The western Pacific is the most active area in the world for tropical cyclones. On average some 30 tropical cyclones form each year with sustained maximum wind speeds of 35 knots, with 19 reaching hurricane status (usually referred to as typhoons) with maximum sustained wind speeds of greater than 64 knots. Many of these cross the Philippines (in 1993 the islands were hit by a record 32 tropical storms) before heading across the China Sea in to the Chinese mainland or veering northwards towards Japan. The incidence of these storms has shown a decline between the late 1950s and the mid-1970s from around 32 to 28, but has since returned to over 30 a year in the 1990s. The incidence of typhoons exhibited a similar dip and rebound.

The other feature of the western North Pacific is that it produces the most intense tropical cyclones. The lowest value of sea level pressure ever recorded was 870 mb at the centre of Typhoon *Tip* on 12 October 1979 (see Fig. 2.17). This intense storm, which lived out most of its existence over the ocean, is a measure of the awesome power of these Pacific tropical cyclones. Only a short way behind *Tip* was the long-lived Typhoon *Gay* in 1992 with sustained wind speeds of 160 knots - a place to steer well away from!

The other important region for tropical cyclones in the northern hemisphere is the northern Indian Ocean, including both the Bay of Bengal and the Arabian Sea. On average, this region experiences just over six tropical cyclones with sustained maximum wind speeds of 50 knots or greater. The most damaging of these run up the Bay of Bengal towards Bangladesh where they can cause immense loss of life, as occurred in 1970 and again in April 1991. This awful loss of life reflects the fact that 40 per cent of the flood plain of the Brahmaputra, Ganges and Meghina rivers is a metre or less above sea level, and so is acutely vulnerable to storm surges associated with tropical cyclones. But the combination of population pressure to exploit the rich soil deposited in the delta regions and rising sea levels (see Section 6.5), means the risk of loss of life will continue to increase.

There is no clear trend in the incidence of cyclones in the North Indian Ocean. This, in part, reflects the low numbers involved and that the storms sometimes follow a wide range of paths from heading westwards across the Arabian Sea towards the Yemen and Somalia, or up towards Pakistan. On average there are one or two cyclones developing each year in the Arabian Sea, usually off the west coast of India. One year in three is normally cyclone-free. One in three cyclones approach the Arabian coast but only rarely enter the Gulf of Aden or the Gulf of Oman. Around the Bay of Bengal they are more frequent and may make landfall anywhere from Sri Lanka to the coast of Myanmar (Burma). Although they can do immense damage in low-lying areas, in general these tropical cyclones lack the extreme wind speeds of the most vigorous systems in the Pacific or Atlantic basin. In part, this may be due to the fact that the season is broken into two by the summer monsoon and so does not peak in the same way as in the other two oceans.

In the southern hemisphere the incidence of tropical cyclones is less than half the total in the northern hemisphere. The principal regions of activity are the south-west and south-east Indian Ocean and the western South Pacific. In all these areas the season is centred on January to March. The south-east Indian Ocean is busiest with an average of 10 tropical cyclones with sustained maximum wind speeds of 50 knots or more. These storms normally track south-westwards initially, parallel to the coast of NW Australia. Many of them later curve towards the south and south-east, making landfall on the sparsely populated coastline and bringing very heavy rains. Due to the extremely limited development on the north and north-west coasts of Australia the high winds associated with cyclones at landfall normally cause very little damage. A notable exception to this was when cyclone *Tracy* destroyed or damaged almost every building in Darwin on Christmas Day in 1974. Tropical cyclones also develop occasionally in the Gulf of Carpentaria on the north coast of Australia. In the south-west Indian Ocean, in an average year eight cyclones will track south-westwards towards Madagascar and Mauritius.

To the east of Australia in the western South Pacific, some six storms a year follow a variety of tracks, ranging from heading south to hit the coast of Queensland, to moving eastwards around Vanuatu and Fiji and as far east as 155°W. The important factor in respect to making landfall in Australia is the state of the ENSO (see Section 2.7). When the sea surface temperatures are above normal in the eastern equatorial Pacific the chances of landfall are significantly reduced. The peak frequency of cyclone formation is in January in the Coral Sea between the New Hebrides and New Caledonia.

3.10. Fog and Reduced Visibility

Fog forms in air, when the temperature falls below the *dew-point* (the point at which the air becomes saturated with water vapour). Where this saturated air is in contact with the ground, it deposits dew, or where it is away from the surface it condenses into tiny water droplets to form fog. Over land these conditions most often occur in low-lying areas on calm clear nights when the ground cools rapidly through loss of heat by radiation and fog forms in the lowest layers of the atmosphere - this is termed *radiation fog*. Over the sea this form of rapid radiative cooling does not occur because of the greater thermal inertia and the vertical circulation of the water. So most fog at sea forms when warm moist air is transported over much cooler waters, and hence is termed *advection fog*. Alternatively, when very cold air sweeps out over warm waters it can produce shallow layers of fog as the water vapour from the sea condenses (or even freezes to form what is often termed *arctic smoke* [although *evaporation fog* is a more accurate description]). These conditions can, in extreme circumstances lead to appreciable build up of ice on ships' superstructures (see Section 3.4).

Because of the rather different meteorological conditions involved with fog at sea, it is important that mariners recognise these differences. Whereas radiation fog is readily dispersed, or lifts into low cloud, when

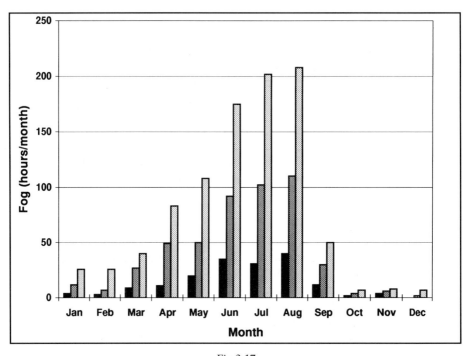

Fig 3.17
Frequency of fog at Nemuro, Japan, (43° 40' N, 145° 35' E) for visibilities of less than 200m
(dark bars), 500m (medium bars), and 1000m (light bars).

there is even a light breeze, at sea advection fogs can occur in strong winds and form to a considerable depth. In extreme circumstances fog can occur with winds of 35 knots and can rise to a considerable height. This occurs from time to time, for example, in the northern North Sea and around the Aleutian Islands. In the latter area fog is known to extend to a height of 1500 metres (5000 feet) at times. Extensive dangerous fog occurs most frequently where cold polar currents extend to relatively low latitudes and come close to warm currents moving polewards. The best known examples are where the Kamchatka and Kuro Shio currents pass one another in the north-west Pacific (Fig. 3.17), or the Grand Banks off Newfoundland at the confluence of the Gulf Stream and the Labrador Current. Similar conditions are found off the coast of Namibia and also off the coast of northern Chile and Peru. More localised examples of these conditions occur where cold deep waters rise to the surface, as off the coast of California around San Francisco. They are also a problem around the Mississippi delta where cold freshwater in winter and spring flowing into the warm Gulf of Mexico produces a sharp temperature gradient (see Section 3.5).

The other significant difference between land and sea fogs is their timing. Radiation fogs are most common during the winter half of the year, while advection fogs at sea are a summer phenomenon. Indeed, in the example given of Nemuro, the concentration in the summer is even more striking, as in the winter (December to February) over three-quarters of the cases of low visibility are due to heavy snowfall. The region between Hokkaido and Kamchatka has the highest incidence of reduced visibility (< 1000 metres) in summer with a figure of around 35 per cent. Around the Grand Banks the figure is about 30 per cent, while between the Aleutians and western Aleutians the figure is about 25 per cent. There are many other local examples of where warm moist air passes over colder water generating dense fog, such as the North Sea haar.

3.11. Terrain Effects

At various points in this chapter the effect of the surrounding land masses has been mentioned as being an important factor in establishing the weather conditions at sea (e.g. the Bora and the Mistral in Section 3.5, and cyclogenesis in the Skaggerak in Section 3.7). These large scale effects are only part of the story. For as inshore sailors know only too well, how the land, sea and tides interact is an essential part of understanding local weather (e.g. sea breezes). Most of these phenomena are adequately dealt with in standard books on meteorology (see Bibliography), but there are some aspects of how islands and larger land features interact with wider weather patterns which deserve separate mention.

Perhaps the most striking terrain effects relate to wind flows around sizeable islands. Good examples can be found from New Zealand to the Canaries, and Corsica and Sardinia. Because winds are diverted round high mountains or forced to flow over lower hills, they establish complex circulation patterns downstream (Fig. 3.18). Even more important is that during stormy conditions the wind funnelling between islands (e.g. Corsica and Sardinia, or the North and South Island of New Zealand) can increase to much greater values than in open waters. These conditions are not usually shown on weather maps and can only be anticipated by having some knowledge of the conditions in the region. In these circumstances the best approach is to consult local forecasters (see Section 5.3). So the simple rule is that when sailing in localities where the adjacent terrain is high enough to exert an appreciable impact on potentially damaging weather systems: check the local forecasts.

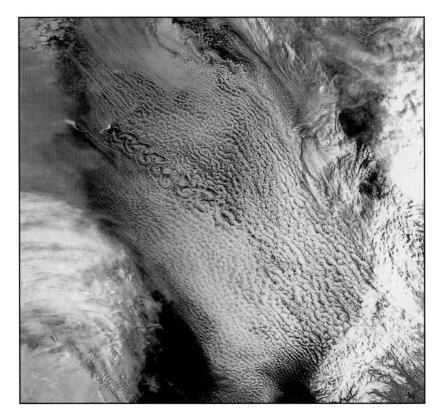

Fig 3.18
A satellite image of vortices formed in the lee of Jan Mayen island showing how the flow of cold air from Greenland in the top left-hand corner of the image is disturbed over much of the distance to Norway in the bottom right-hand corner (with permission of the University of Dundee).

FOOTNOTES

[1]   See US Navy (1977), and other global climatologies published by national authorities and universities.

[2]   The US Navy Marine Climatic Atlas of the World is available as a CD ROM (price US $120) from the National Climatic Data Center, Ashville, North Carolina.

[3]   Young & Holland (1996).

[4]   The KMNI work is Korevaar (1990). Other examples of these types of regional analysis include Draper (1991) and Isemer & Lutz (1985),while detailed analyses of the Adriatic, ice conditions in the Baltic and hurricanes in the northwest Pacific can all be found in the National Meteorological Library at Bracknell, England.

[5]   Boutin & Etcheto (1996).

[6]   Barstow & Lygre (1985).

[7]   Bacon & Carter (1991).

[8]   Hulme & Jones (1991) provides a detailed description of the challenges of producing a reliable measure of gales over the British Isles. Seasonal figures for average windiness in the southern North Sea appear in WMO (1995), p 78.

[9]   IPCC (1995), p. 166.

[10] Trenberth & Hurrell (1994).

[11] Torrance (1995).

[12] UK Goverment (1969).

[13] Shellard (1974), which draws on earlier work by Mertins (1968).

[14] UK Admiralty, Arctic Pilot.

[15] This breakdown can occur up to a month earlier. In September 1969 the explosive development of a depression to the south east of Malta on the 23rd producing gale force winds in Malta which sank the 20,000 ton tanker *Angel Gabriel*. It then moved south-westwards causing unprecedented floods in North Africa which killed over 600 people in Tunisia and northeastern Algeria.

[16] Cardonne *et al.* (1996).

[17] Hsu (1993).

[18] This generally accepted view was called into question in April 1991 when a tropical cyclone formed off Angola. This is the only documented tropical cyclone in the South Atlantic although it is possible that there have been others, in the days before satellite imagery. On the other hand, it could be a genuine "first" resulting from global warming.

[19] UK Admiralty (1941), page 369.

[20] Landsea (1993).

[21] Crutcher & Quayle (1974).

CHAPTER 4

WEATHER FORECASTING

"Time present and time past
Are both perhaps present in time future
And time future contained in time past."

T. S. Eliot, 1888-1965

Two complementary features of weather forecasting are central to getting the most out of what is produced by national weather services around the world. The first is the huge scale of the analytical process involved producing the forecasts. The second is the immense complexity of global weather and the inherent challenges facing forecasters in handling what is essentially an unpredictable system. This involves a balancing act between exploiting what is an increasingly high quality product while recognising the limitations in what can be forecast. So the objective of this chapter is to show that, whatever your interest in maritime weather and climate, it is possible to get a great deal more value from forecasts by knowing about the thinking behind them, without going into too much technical detail.

4.1. Numerical Weather Prediction

The key to exploiting forecasts is a basic understanding of what is involved in their production and what this can provide in predicting the future behaviour of the atmosphere. The science of weather forecasting today is dominated by the use of huge computers. The advances in this technology have been central in the development of numerical weather forecasting over the last 40 years. Computer models of the global weather treat forecasting as a problem in mathematical physics. They contain a very large and complex array of equations based on the physical and dynamical laws which govern the birth, growth, decay and movement of weather systems. This involves not only modelling as many of the physical processes as possible, but also representing relevant aspects of the atmosphere, the oceans, and the Earth's surface. This is an immense area and requires a large number of approximations and simplifying assumptions to make the models manageable.

To predict the weather several days ahead these models have to consider the entire global atmosphere and compute its future state at regular intervals. The physical state of the atmosphere is updated every six hours using observations from around the world from surface land stations, shipboard equipment, buoys, and in the upper atmosphere using instruments on aircraft, balloons and satellites. In the most advanced models the atmosphere is divided in 30 layers, and each level is divided into a network of grid-points about 60 kilometres apart. So the atmosphere is represented by some four million points, and is rapidly becoming the most demanding part of the forecasting process in terms of computer power, as ever-increasing amounts of data become available, especially from satellites. Once the process of initialisation has been completed, each of these points is assigned new values of temperature, pressure, humidity, and wind speed. The basic process of assimilating all the available data and producing an estimate of the current state of the atmosphere already takes about a quarter of the computational effort and this proportion is rising fast as more and more data become available from satellites and other measurement systems. Then the model is run to calculate the movement of the atmosphere in 15-minute steps. In all, predicting the future state of the atmosphere up to 10 days ahead requires several tens of trillions calculations.

Because this forecasting process calculates a step-by-step analysis of future global weather patterns throughout the depth of the atmosphere, it produces a wide variety of outputs. What is more, the best forecasts (e.g. the product of the European Centre for Medium-Range Weather Forecasts) can be used to predict the weather anywhere from the South China Sea to the North Sea. While the accuracy of the predictions is likely to vary from place to place, depending on the weather situation, the global model provides forecasts for any part of the world. The forecasts from many national weather services use a limited number of similar models and are capable of achieving comparable results. Much of the success depends on how the predictions are interpreted by forecasters and used by mariners.

Normally, the standard product which reaches the end-user presents pressure patterns, with superimposed weather features (e.g. surface fronts), at different levels in the atmosphere, but many other forms of output can be produced. These products are largely defined by the output of computer models, but in a variety of important ways have already been the subject of interpretation and refinement by forecasters. In particular, the identification of fronts and their associated weather still needs an element of human judgement. For mariners understanding this combination of computer output and forecasters' analysis is central to taking full advantage of any predictions. At the most basic level the predictions of weather patterns and surface wind speeds are what matters most, especially when combined with calculations of the sea state that will be generated by the predicted wind fields (see Section 4.4). These are effectively generated by the computer. But the interpretation of regions of active weather may be assisted considerably by discussion with expert services.

The value of expert guidance increases further when there is a need to go into more detail. The skills of local forecasters can add appreciably to the analysis of the global predictions produced by major forecast services. Their contribution relates to two principal areas. The first is in the use of limited area forecasts which are produced by similar numerical prediction methods for a smaller part of the world in greater detail. These forecasts assume that at the boundaries of their domain conditions are represented by the output of the global model and are only used for shorter periods ahead (up to 72 hours). They are produced by the major forecasting organisations for various parts of the world and can compute weather systems with a resolution of down to under 20 kilometres. The principal benefit of these forecasts is to provide more precise analysis of the structure and timing of particular weather features (e.g. localised areas of heavy precipitation and strong winds). As such they provide local forecasters with a better picture of how the weather is developing in their particular locality, which enables them to provide a more detailed service for users.

The second refinement is the ability of local forecasters to provide interpretation of the output of the computer models in terms of their knowledge of conditions in their part of the world. These insights are an essential part of making effective use of numerical weather prediction, because, while the broad features of the forecast are instantly recognisable the detailed aspects may only be apparent to the expert, often with the benefit of local knowledge. The human forecaster can add value in two ways. The first is to make an assessment, usually subjectively, of the probable limits of error in the numerical output and to incorporate this into any forecasts issued. This may often involve what is termed "pattern recognition" which draws on the experience of the forecaster to anticipate locally dangerous weather. In coastal areas and, in particular, when forecasts are required for a port or some other inshore location, the human forecaster can add a great deal to the output from numerical models, especially if he has good local knowledge of the location. The essence of this process is that the present generation of numerical models provide guidance material for use by meteorologists rather than forecasts for use by end users. This means that services providing forecasts for limited areas can draw on a wide range of computer outputs and combine them with local knowledge. Against this background, it is foolish for mariners to accept the predictions of any one model at face value, and rely on these alone to make decisions.

This in-depth analysis relies on not only the expertise of the local forecasters, but also the use of information which may not be fully exploited by the computer forecasts. This includes satellite images (see Section 4.3) and weather reports which have become available since the computer forecasts were set in motion. This type of refinement is of particular importance where weather systems are developing rapidly. This is an area where the computer models tend to underestimate the pace at which events can unfold. In these circumstances local interpretation can combine the latest incoming observations and extend to individuals reading the signs in their vicinity. But, in deciding that weather may not be developing in quite the way predicted, it is vital not to overlook the scale of the effort that has gone into the production of the standard forecasts. So before trying to second guess the forecasters, mariners must ensure they fully understand the latest predictions and also have confirmed with local services whether there are any more recent important developments. Only then can they begin to think seriously about putting their own gloss on what is happening about them. In so doing, it helps to know a little about the progress being made in forecasting and the strengths and weaknesses of current predictions.

4.2. Forecasting Performance

Progress in weather forecasting is the case of some good news and some bad news. The good news is that the growth in computing power, the widening understanding of the physical processes involved, and improved measurement technologies are producing better forecasts. The bad news is, that with the progress comes the growing awareness of the immense complexity of the weather which imposes real limits on what can be forecast, especially in terms of timescale of predictions.

Broadly speaking, it is now possible to produce useful forecasts, which show some degree of skill as compared with relying on the weather continuing in its current pattern or simply drawing on the standard climatology of the region, out to about five days ahead on average. By comparison forecasts in the early 1970s could only achieve the same results two days ahead, although during the 1990s the rate of progress has slowed. At the same time, the inherent limitations in predicting atmospheric behaviour, because errors in the knowledge of the current state of the atmosphere double in the modelling process every two days or so, have become more apparent. This means that useful forecasts of the day-to-day weather beyond about 10 to 14 days are impossible. Also it is becoming increasingly clear that performance varies significantly from time to time depending on the overall state of the atmosphere.

Another problem is that while a model may do a good job of predicting a major development on the synoptic scale, less eye-catching aspects may not be well predicted. This is important as most users of forecasts are concerned only with a single location at any point in time, and so the ability to get certain features right may be of only limited relevance if other details are less well predicted. This localised response is best considered in terms of forecasts of extreme events. A good example of this was the prediction of the record-breaking low in December 1986, which deepened to 916 mb in the vicinity of Iceland. This depression ranks alongside the Braer Storm (see Section 3.2 and Fig. 3.4) as the most intense ever recorded in the North Atlantic. Although the UK Meteorological Office model predicted the intensity of the storm from 3 days ahead with considerable accuracy,[1] its position was initially expected a little farther east, and as a consequence, the forecast for the pressure gradient over the North Sea was expected to be much stronger. So while anyone sailing to the south of Iceland would have been justifiably impressed by the forecast this extreme situation, operators in the North Sea would have been preparing for much stormier conditions, which did not arrive as predicted.

The more general property of how errors grow in numerical weather prediction could, however, be the key to improving forecasts. Starting from similar initial conditions the speed with which differences in the forecasts grow provides a valuable insight into the predictability of current weather patterns. This is related to whether or not the circulation patterns are stuck in a stable regime. As was discussed in Section 2.4, these blocked conditions are most important in the mid-latitudes of the northern hemisphere in winter, when three quarters of the time the weather is stuck in one of four or five patterns which can persist several days or longer. During these quasi-stationary periods the weather behaves in a more predictable manner and useful forecasts well beyond five days may be possible, although this may only become apparent after the event. But, when the regime breaks down, the change is rapid and often unexpected and the forecasts deteriorate much more rapidly. Similarly, when the atmosphere does not settle into a stable regime for a while the forecasts are less reliable.

Weather forecasters are confronting the challenge of varying predictability of the atmosphere by using a statistical approach, known as *ensemble forecasting*.[2] The underlying assumption in this form of forecasting is that if the model is run with slightly different initial conditions, then depending on whether the atmosphere is in a predictable or unpredictable mood, the errors in the predictions will grow at different rates. If the atmosphere is stuck in a given regime then the errors will grow slowly, so a set of forecasts starting with tiny differences in their initial conditions will diverge only slowly over time. If, however, the atmosphere is less stable then the same tiny differences will lead to rapidly diverging predictions. So, depending on whether the ensemble of forecasts stick together for a week or more, or rapidly head off in different directions, the forecasters know when to put greater or lesser trust in the output of the models.

The availability of an ensemble of forecasts has a variety of implications for the users of forecasts, over and beyond just knowing if the weather is in a predictable or unpredictable mood. The differences that show up in the ensemble may also provide insights into whether there is a significant chance of conditions developing in one or other of two very different directions. Alternatively, the ensemble may lean principally in one direction but identify a lower chance of a more dangerous condition. These alternatives need to be considered by operators who are embarking on risky weather-dependent activities, and can only be fully appreciated by consultation with those involved in the forecasting process.

A good example of applying this statistical approach to medium-range forecasts is in ship-routeing (see Section 4.6). For instance, crossing the North Atlantic or North Pacific involves voyage times comparable to the timescale of the forecasts. Furthermore, the improved predictions produce greater benefits in winter when the risks of delays due to adverse weather are most likely, and when the influence of stable regimes is greatest. At the simplest level, if the forecasters have to admit that the models are unreliable beyond a few days then operators may be advised to rely more on climatology than the uncertain future presented in the forecasts. More interestingly, forecasters may identify a situation where the most probable developments point to following one route, but where there is, say, a one-in-five chance of a severe storm developing late in the forecast period, whereas taking the alternative route, although slightly more expensive, greatly reduces the risk of being caught by such an event.

These uncertainties often involve switches between different weather regimes. The existence of such regimes are most easily visualised in the pressure and wind patterns in the middle troposphere (see Section 2.2). Standard computer output presents these patterns in the form of maps of the height of the 500-mb level, which corresponds to an altitude of about 5500 metres (18 000 feet). Although these maps bear a marked similarity to surface pressure maps, there is an important difference in that they show contours of the height of the 500 mb surface above mean

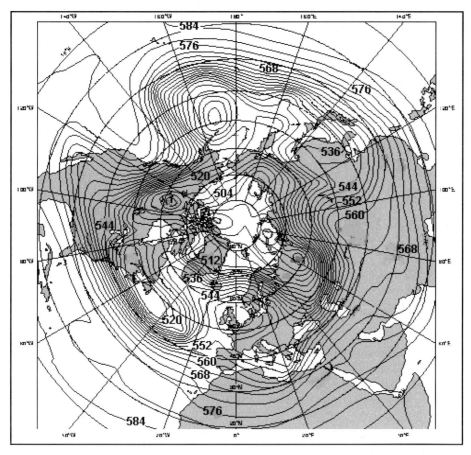

Fig 4.1
A map of the 500-mb level issued by the ECMWF on 2 January 1997 showing the prediction of the pattern for 8 January 1997. The important feature of this output is the meandering circulation pattern with distinct blocking over the British Isles and off the west coast of North America. (with permission of ECMWF.)

sea level. Although intended primarily as a tool for use by professional meteorologists they do give the non-expert some indication of the main broad-scale weather features. A 500 mb low will usually have an accompanying surface low but with a centre which is displaced horizontally from the centre of the 500 mb low. Similarly, a 500 mb high will usually have an accompanying surface high, again with some horizontal displacement of the centre. Where the 500 mb contours are close together it is likely that there will be considerable activity at the surface with mobile frontal systems and the possibility of fast-moving and rapidly developing low pressure systems. In general, in middle latitudes, closely spaced 500 mb contours suggest very disturbed and often windy conditions at the surface. Widely spaced 500 mb contours suggest much quieter conditions at the surface.

In interpreting 500-mb charts the important point to remember is that many of the detailed features of surface weather are absent. Instead, the contours provide a clear picture of the main features of the circulation in the mid layers of the atmosphere. The zones of the steepest contour gradients between the major areas of high and low pressure coincide with the strongest winds. These occur at somewhat higher levels, where the core of the winds is known as the *jet stream*. The jet stream is normally some thousands of kilometres in length, a few hundred kilometres in width, and some kilometres in depth. The wind speed decreases rapidly above and below the jet stream axis, and on either side of it. In the case of the *polar front jet stream* (see Fig. 2.4), the greatest wind speeds are usually around 200-300 mb between latitudes 40° and 60° in association with polar front depressions at the surface.

The importance of these strong upper atmosphere winds is that they steer the surface weather systems and so maps of their predicted behaviour help to envisage the movement of weather systems. These are much more serpentine affairs than the examples shown in Figure 3.3. This can be seen in the example of a forecast prepared six days in advance by the ECMWF which is shown in Figure 4.1. At the time there was a block over the British Isles and off the west coast of the USA. These features were predicted to continue with a strong westerly jet around 40°N steering depressions across the Atlantic into the Mediterranean. At the same time, a second flow would divert depressions up towards the Davis Strait. In the Pacific the pattern is pushing storms along a southerly track until 150°W and then sweeping them northward to Alaska.

This type of presentation makes it much easier to see why weather systems are following unusual courses. Furthermore, if, as was the case in early January 1997, the pattern is stuck in this blocked weather regime, and this is predicted to continue over the forecast period then the wayward movement of surface weather features may well prove to be close to the mark. When, however, the upper winds are predicted to be in a state of flux, a state often associated with a fast-moving westerly pattern, with both their wavelength and amplitude varying, then the forecasters task is more daunting. At such times the user should pay particular attention to the uncertainties expressed in any forecast concerning the position and intensity of storm systems.

Another aspect of the upper atmosphere circulation worth mentioning is how the position and intensity of the waves are part and parcel of the large scale circulations at the surface. This is because of how air converges and diverges as it flows through the waves. *Divergence* occurs ahead (to the east) of a *trough* in the upper westerlies, inducing *convergence* at the surface, with rising motion in between. This means that the region beneath the eastern side of a trough is a favoured area for depressions, deep cloud formation, and precipitation (Fig. 4.2). *Convergence* occurs ahead (to the east) of a *ridge* in the upper flow, inducing *divergence* at the surface, and subsidence between the two levels. This means this region is favoured for the development of surface anticyclones, or ridges of high pressure, and dry, clear conditions,

There is an important underlying message in these observations about weather systems and the upper atmosphere. Many mariners are inclined to consider the weather in a two-dimensional way, concentrating solely on surface pressure patterns. As explained in Chapter 2, meteorologists consider the atmosphere in three-dimensions with vertical motion being an essential part of any analysis. In so doing, they get a better understanding of the physics behind the predicted developments in the patterns at the surface. In terms of any upper air analysis, the mariner can also benefit by getting a sense of how things are developing throughout the depth of the atmosphere even though any predictions have to be taken at face value. But when combined with discussions with the forecasting service it may be easier to exploit their products to good effect.

At the more detailed level, the increased resolution of both global computer models and limited area models has improved the definition of developing weather systems. But the models still have difficulty with the most rapidly developing depressions (*explosive cyclogenesis* - see Section 2.5) which represent such a major hazard to shipping. Depending on how fast they develop and move across the ocean, not only do they generate excessively steep seas, but they also exchange huge amounts of heat, moisture and momentum which can have a significant impact on how

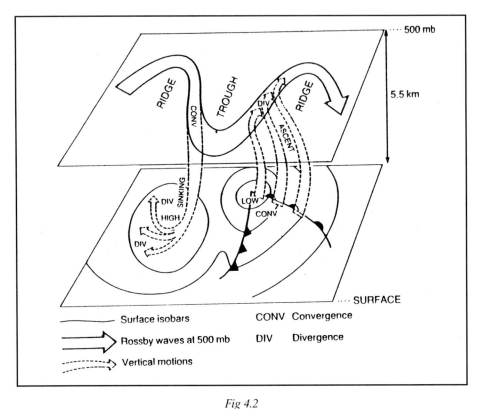

Fig 4.2
A schematic representation of the relationship between the location of surface highs and lows and the troughs and ridges in the upper atmosphere (Rossby) waves (from Musk, 1988. with permission of Cambridge University Press.)

quickly they deepen. In general, the models tend to underestimate the rate of deepening of the most explosive systems, even though the consequence of the roughness of the sea state is to slow the winds down a bit and reduce the pressure drop during the early stages of development (see Section 4.5). A complementary weakness is that they tend to underestimate the rate at which the depression fills, although this more rapid return to normal is less likely to pose dangers to shipping. These shortcomings also cause longer term problems for forecasts as the failure to predict the development of intense depressions has knock-on effects for the subsequent output of the computer model.

The challenge of rapidly deepening systems over the oceans is also reflected in predicting their movement. Often there are not adequate data to pick up their early development and speed of movement. This makes the precise forecasting of the paths of both extratropical depressions and tropical cyclones still a long way off. So it is wise to steer clear of rapidly deepening systems on the basis that the forecast track could well be wide of the mark. Broadly speaking, the *average* errors for the prediction of the central position of tropical cyclones are approximately 100 nautical miles (185 km) for 24 hours ahead, 200 nautical miles (370 km) at 48-hours, and 300 nautical miles (550 km) at 72-hours. The errors in predicting the tracks of individual storms can be much greater. Despite the improvements in numerical modelling in recent years there has been little corresponding improvement in these error statistics. Similar statistics for extratropical depressions are more difficult to present because of how systems are linked to one another and they evolve over time.

These uncertainties have significant implications for mariners using forecasts. In most cases a single prediction is made for the track, speed of movement and rate of development of any particular low pressure system. Any forecast issued is normally based on these predictions. The user should always bear in mind that the predictions are likely to be either the direct output of one numerical model, possibly involving an ensemble of estimates, or the distillation of the outputs from two or more models. In the future the predictions will increasingly be the result of using ensemble techniques. No matter how the set of predictions is arrived at, the user should always bear in mind that it represents the most probable scenario out of a range of possibilities. That range may vary appreciably, depending on the inherent predictability of developments in the atmosphere at the time. In most cases, forecasts do not include any information about the range of possible errors either side of what is deemed as being the most probable.

Things are, however, starting to change. In some parts of the world, notably the North Atlantic, probability forecasting is now beginning to be used with tropical cyclones. Basic forecasts are still prepared on the assumption that the cyclone will follow the predicted track with the predicted speed of movement and rate of development. These forecasts are then complemented by predictions of the probability of the centre of the cyclone passing within 65 nautical miles (120 km) of a selection of each of a number of towns and cities on land and spot locations at sea. Taken together these forecasts tell users what is considered to be the most probable evolution of the weather plus an estimate of the uncertainty inherent in the prediction. A similar approach to extratropical depressions is less likely to be developed as these systems are much less well-defined than tropical cyclones. Furthermore, their direct effects at any particular point are less directly related to the detailed movement and development of the centre.

All of this reinforces the basic message that it helps to have an informed view of the limitations of any forecast. This does not mean that mariners should try and do their own forecasting. Conversely, they should not fall into the trap of misinterpreting the confidence of forecasters as meaning that the charts they use as part of their predictions can be regarded as actuals. What they need to do is develop the skill to ask the right questions about what they are getting. There are times when it is easier to exercise this judgement. For instance, when the atmosphere is blocked forecasts are likely to be more accurate than when there is a lot of mobility taking place. This is a simple example of combining the physical understanding implicit in models and the huge power of supercomputers with their own operational experience including knowledge of the aberrant behaviour of the weather. In this process numerical model output should not be treated as an end product in itself when dealing with weather-sensitive marine operations, but as a part of the armoury of experienced marine weather forecasters in advising mariners on how best to anticipate weather developments. As the various examples in Chapter 3 show, the forecasters often have the greatest difficulty with the most vigorous systems, when the most extreme sea states will occur. This means that developing the ability to understand the strengths and weaknesses of their services may be of vital importance when mariners run the risk of encountering the most vicious conditions.

4.3. Weather Satellites

Weather satellites, as opposed to the oceanographic research satellites discussed in Sections 2.5 and 3.2, are now an integral part of both meteorological science and weather forecasting.[3] The images they produce are part of our lives, so it is important to know something about what they can do. But, in reviewing their role in meteorology, it is wise to start with a cautionary word. Satellite images contain a great deal of information which, in the right hands, can provide valuable additional insights. If, however, they are used selectively by the uninitiated to reinterpret standard forecasts, they have the potential to mislead. So, as a general rule, satellite imagery is not something which ships' officers should be using to override the forecasts from standard sources. What a background knowledge about weather satellites can do is provide a better understanding of how measurements from space are utilised to produce weather forecasts and more generally the dynamics of the Earth's atmosphere.

The most obvious advantage of polar-orbiting satellites (currently NOAA 14 and 15), is that they provide effectively global coverage of weather systems every 12 hours. At the same time a ring of five geostationary satellites sitting above the equator at a distance of some 36,000 km (including GOES W at 140°W and GOES E at 75°W, Meteosat over the Greenwich meridian plus the Indian satellite (INSAT) at 70°E and the Japanese satellite (GMS) at 140°E), provide images every half hour to latitudes around 70°N and S. This combination is particularly valuable for monitoring conditions over the oceans where insufficient regular measurements are made, and so is of vital interest to mariners (see Figs. 2.5 and 2.17).

At the simplest level weather satellites take a series of images of atmospheric features using both visible and infra-red wavelengths. The visible wavelengths are sensitive to all forms of cloud and so provide a clear picture of the broad features of weather systems, but cannot detect features deeper down which are often obscured by thin high cirrus. Also they can only make observations during daytime. Infra-red images usually measure the temperature and are better able to peer into weather systems and pick out the thickest areas of cloud. They do, however, have difficulty in detecting low cloud and fog which is at roughly the same temperature as the surface.

In addition, the satellites are equipped with radiometers which are able to measure the temperature profile of the atmosphere. These instruments measure both infra-red and microwave heat radiation welling up from the atmosphere and the surface below (see Section 2.1). The infra-red devices are capable of measuring the temperature at four broad levels in the atmosphere. The observations are, however, contaminated by radiation from clouds which restricts their value. Microwave observations are not affected by clouds but can only discriminate between two levels in the atmosphere which also limits their value for forecasting work. In addition, these instruments can make accurate measurements of sea surface temperatures (Fig. 4.3) and the extent of pack ice (see Figs. 3.7 and 3.8).

All of this information can be utilised by forecasters to improve their predictions. The images of cloud cover are used to check whether the computer models are reproducing the current state of the atmosphere or whether there are tell-tale signs of rapidly emerging developments which are not picked up in forecasts. These come in many shapes and sizes and have been the subject of detailed analysis in standard predictions.[4] The clear message is that in the hands of skilled forecasters these images can provide vital clues about unexpected weather developments (Fig. 4.4), but for the rest of us they should make forecasts more accessible without tempting us to jump to rash conclusions about being able to reassess the predictions.

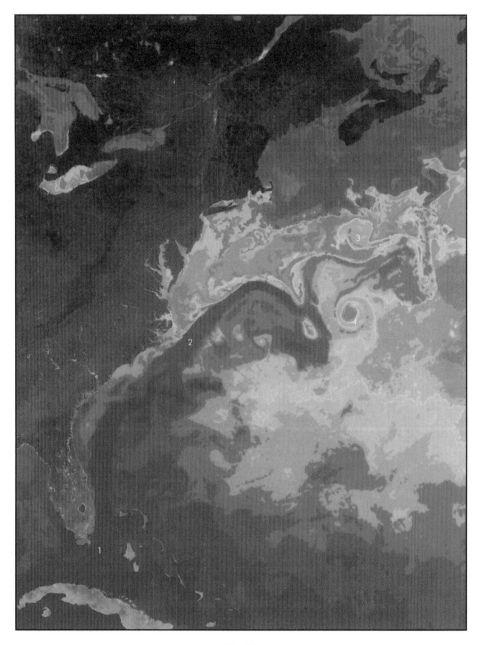

Fig 4.3
Temperature measurements of the Gulf Stream in April 1984 obtained from 35 satellite passes showing the warm water (1) sweeping up from the Gulf of Mexico and out across the North Atlantic (2). At the same time, cold water sweeps down from Newfoundland to Cape Cod. Of particular interest is how the Gulf Stream tends to break up forming warm (3) and cold (4) eddies (with permission of NASA).

Temperature measurements are incorporated directly into the NWP models.   But given they are blurred out over such thick slices of the atmosphere, this process is not simple.   In effect the models, instead of using the temperatures obtained from the satellites, calculate how much radiation they would expect to see welling up from the atmosphere, and where this differs appreciably from what is observed, adjust the temperature fields to bring the model into line with satellite measurements. Other data from satellites (e.g. detection of low cloud and fog, or the extent of sea ice) can be used directly by forecasting services to assist in providing warnings of adverse conditions.

Research satellites are producing a range of oceanographic data which has substantial potential for improved forecasts in the future.

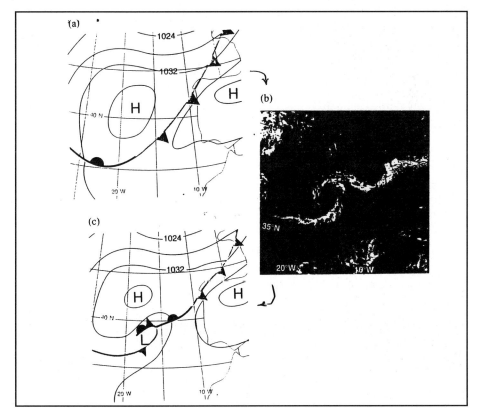

Fig 4.4
An example of how satellite images can identify developing features which sometimes do not show up on the surface analysis, where (a) is the preliminary surface analysis at 0600 UTC on 4 January 1989, based on observational data, (b) is the visible satellite image at 1438 UTC, and (c) is the amended surface analysis for 0600 UTC taking account of the satellite information. (From Bader et al. 1995, with permission of Cambridge University Press.)

Work on measuring wind and waves, which has already been discussed in Chapter 3, will be considered in the next section. More esoteric studies of the height and strength of the major ocean currents, and the propagation of circulation patterns with a characteristic wavelength of hundreds of kilometres, are still in the realms of research. But, as we will see, in considering the prediction of ocean currents (see Section 4.7), and forecasts of ocean-atmosphere interactions (see Section 6.3), these studies may well hold the key to an improved understanding of ocean dynamics, and, hence better forecasts in due course. Another area where satellites are making a significant contribution is in measuring global climate trends since the 1960s, and this work is reviewed in Chapter 6.

4.4. Forecasting Waves

The prediction of wave heights and directions across the world's oceans is an area of forecasting that has advanced significantly in the last couple of decades. This progress is the product of both a better physical understanding of the process of energy exchange between wind and waves, and the exploitation of new technologies, including supercomputers, automatic buoys and satellite remote sensing. By the late 1980s the models were able to do a reasonable job of forecasting most sea states, using the wind fields produced by the standard NWP models, several days ahead, but there were two notable exceptions.[5] These were a tendency to underestimate extreme sea states, and an underestimate of wave height by about 20 per cent in large parts of the Southern Hemisphere and tropical regions during the austral winter, which appeared to be linked to the prediction of swell from storms over the southern oceans dissipating too rapidly (see Section 2.5).

The latest models, of which the ECMWF version is a good example, have incorporated a number of important refinements. First, they exploit the increased spatial resolution to provide a more detailed representation of surface winds which should improve the analysis and prediction of waves in storm tracks. At the same time the horizontal resolution of the wave field has been doubled from $3°$ to $1·5°$. Comparison of forecasts with buoy observations confirms that the better representation of the wind field improved the prediction of both wave heights and extreme sea states.

The second advance is the more realistic treatment of the generation of waves and the consequent slowing down of the air as momentum is transferred to the sea surface. This leads to waves growing a little more rapidly. It also results in the model predicting a reduced and more realistic level of short, choppy and breaking waves. This reduces the attenuation of swell in the vicinity of storms, which has improved the forecasts in the southern oceans and had a significant impact in the tropics where wave heights are dominated by the swell generated at higher latitudes.

The third advance is to use altimeter data from the ERS 1 and ERS 2 satellites to establish the initial wave conditions in the model. These satellites have had the facility to analyse some of their altimeter observations onboard to calculate significant wave heights and make them available to operational centres within three hours. This additional information gets the forecast off to a better start and improves the predicted wave heights and direction up to five days ahead. In part this is a result of the satellite data providing better values of the underlying swell in the wave field.

Overall, checks made by using available buoy observations show that the current ECMWF model slightly underestimates the average wave height by 20 cm with a root mean square error of 50 cm. The best results were obtained around Hawaii, in the eastern North Pacific and the North East Atlantic. For the western North Pacific the results are less good and relatively poor off the east coast of the USA. The latter may be linked to using buoys moored in the Gulf Stream, where the combination of the strong current and frequent local cyclogenesis, can produce extreme wave conditions (see Sections 2.5 and 3.2), and these appear to represent a sterner test for the wave models. As for the performance over time of the forecasts, in the northern hemisphere they produce useful predictions up to five days ahead. In the southern hemisphere, with its less marked seasonal cycle, they deteriorate more quickly and are currently of little value beyond four and a half days, while in the tropics where the variability is much less, their utility extends to seven days.

The performance of wave models in handling extreme sea states has been examined in terms of two storms off the east coast of the USA (the Halloween Storm in October 1991 and the Storm of the Century in March 1993 (see Section 3.6).[6] This study showed that the various models performed well in retrospectively calculating sea states using observed wind speeds from buoy data with significant wave heights up to 12 m (40 feet), but systematically underestimated the peak sea states for both storms which exceeded this value. While the most advanced models did slightly better in this respect, none of them could reproduce the wave generation along a dynamic fetch associated with intense wind maxima, termed *jet streaks*. These cores of exceptional winds with central speeds of around 60 knots appear to be an integral part of rapidly-deepening storms. They maintain high spatial coherency for at least 24 hours and travel within the storm system but tend to move somewhat faster than the centre of the low progressing at speeds of around 30 to 40 knots.

   This phenomenon of localised cores of extreme winds may explain the erratic distribution of damage in other well-documented storms (e.g. the Fastnet Race storm in August 1979 [see Section 5.4], the Great October storm that hit south-east England in October 1987, and even the variation of the immense damage caused in Miami by Hurricane *Andrew* in August 1992). The existence of these low level jets has been supported by recently published work[7] which has measured the winds in a hurricane using Doppler radar equipment. These observations were obtained when hurricane *Fran* hit the South Carolina in 1996 and showed intense wind rolls in the atmosphere close to the surface and parallel to the general wind direction which caused the winds speeds to vary over a factor of two across distances of a few hundred metres. This phenomenon appears to explain why hurricane damage can vary so much over short distances.

   These observations also underline the problems of forecasting extreme sea states as it will require much more refined high-resolution models of storm systems if the conditions for forming jet streams are to be identified and their position predicted with any accuracy. But, at the practical level, it confirms that mariners should expect to encounter extreme sea states when the rapid development of a major storm is forecast (the progress of the Storm of the Century was accurately predicted several days in advance), and these can comfortably exceed the figures in the standard climatologies. In effect, the tales of old sea dogs having survived huge waves have more than a grain of truth in them.

   The generation of swell mentioned above is another feature of these massive storms which poses a threat to distant maritime activities. Because swell with a period of 20 seconds travels away from the storm at a speeds of 60 knots (see Section 2.5), it will cross the North Atlantic in two to three days. Models which can predict the generation, propagation and dissipation of these waves are of potential value to offshore operations to the west of the British Isles, because floating platforms are particularly susceptible to long-period waves. The UK Meteorological Office, in conjunction with W. S. Atkins, who have an international reputation in motion-calculation services for offshore platforms, have recently developed a service for providing forecasts of significant and maximum heave up to five days ahead. These calculations of heave take account of the vessel's position, heading and draught, and can cover pitch, roll and sway motions. Initial results suggest that these forecasts are capable of achieving high accuracy, and offer significant cost savings, greater efficiency, and increased confidence in operational management. Wider use of services like this should eventually reduce the type of damage caused by swell from distant storms described in Section 1.2. Increasingly, given the advance warning that be provided, operators will expect harbours that are particularly vulnerable to such conditions to have warning services to reduce the chance of damage.

In concentrating on the broader climatological consequences of forecasting waves and stressing the interest of this work for long ocean voyages, it is all too easy to lose sight of the benefits of detailed shorter term local forecasts. In particular, the increasing operations of high speed ferries require accurate advice on whether wave conditions will exceed certain thresholds, as their operations are often governed by licences issued by national maritime safety authorities. This requires high resolution models with an accurate representation of local bathymetry to be able to provide reliable estimates of sea state on specific routes. There are many other examples of maritime activities with low weather thresholds where the provision of detailed short term forecasts are an essential part of safe and efficient operations.

4.5. Weather Routeing

The benefits of weather forecasting for planning voyages consist of saving time and bunkers, and minimising damage to both the vessel and its cargo on transocean crossings. These have been well recognised by the shipping industry since the 1960s. Since then the advances in weather forecasting, together with the development of models to make accurate forecasts of the waves that will be generated by the predicted wind fields and how this will affect the progress of any vessel, have increased the value of these services. But, before explaining the benefits of weather routeing to both ships' masters, owners and charterers, it helps to set the analysis in the context of the two practical aspects of both shipping operations and producing and using the forecasts.

In a little more detail the purpose of weather routeing may be of use for any of the following:
- to keep fuel costs to a minimum;
- to assist in the vessel keeping to a rigid schedule;
- to prevent damage to the vessel;
- to prevent damage to the cargo; and
- to minimise discomfort to passengers.

Depending on which of these objectives has greatest priority, advice provided by routeing services can be used in different ways. In particular, where commercial considerations put a high price on meeting a tight schedule then speed will be of the essence, whereas luxury cruising may set a far higher store on avoiding any passenger discomfort. These differing demands mean that the services and their customers need to maintain a close dialogue in providing relevant information to assist in making balanced decisions.   In this process, the weather thresholds at which problems begin to occur and the speed of the vessel are two important considerations when exploiting the weather routeing service.

The basic strategy for weather routeing can be done well in advance based purely on climate.   This may indicate that there is likely to be only one feasible basic route.   Alternatively, climate may show that there are two or more feasible routes.   An example of this is whether to go northabout the British Isles or southabout when departing from a Northwestern European port for a transatlantic voyage.   Where there are a number of possible strategic routes a decision can be made at the time of departure based on the latest Numerical Weather Prediction output.

The type of forecast prepared will depend to a large extent on the nature of the voyage.   For the typical long distance cargo operation will consist of a recommended route together with a forecast up to 10 days ahead of:
- wind speed and direction;
- wave height, including a breakdown of how much is generated by swell and how much by the local wind field;
- the direction and period of the swell; and
- visibility and any other significant weather

Usually this information is displayed in either graphical or tabular form (Fig. 4.5).   This combination makes it possible for the ship's master to see at a glance the conditions likely to be met on the route.

This leads into the practical problems of using forecasts.   These are the limits of how far ahead predictions can be made and the balance between the advice provided by weather routeing services and the master's responsibility for the safe and efficient management of the ship, which depends on past experience and the ability to read accurately current wind, wave and ship response characteristics.   The limits of forecasting have already been discussed in Section 4.2 and for practical purposes must be regarded as around five days, and may be somewhat less than this when the atmosphere is in a volatile state.   The fact they can do considerably better in some circumstances is only of limited use in maritime operations.   This means that voyages lasting significantly longer than five days or so are bound to face considerable uncertainty about the weather in the later part of their trip.

Even with useful forecasts several days ahead, there may be a good reason for deciding principally on climatology.   For instance, a major oil company operating modern tankers to ship oil from the Shetlands to the

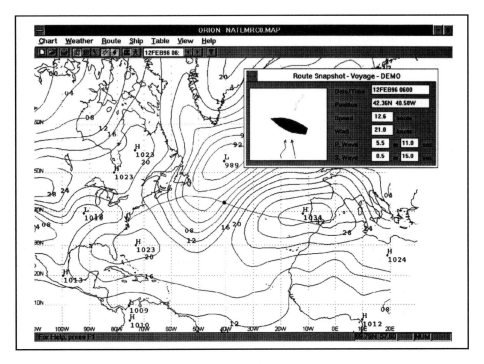

Fig 4.5
An example of the type of advice that is issued by weather routeing services showing advice for a voyage from the Straits of Gibraltar to the East Coast of the USA where the optimum route is a northerly track to minimize exposure to head conditions. (with permission of Oceanroutes.)

East Coast of the USA in winter may be well advised to operate on the assumption that, with a steady stream of depressions moving across the North Atlantic, the best course is to take the most direct route, subject only to ice conditions between Greenland and Newfoundland. Where vessels can handle the heaviest weather and the fluctuations between successive storms tend to average out, the emphasis on the most direct route to achieve the desired speed may prove the right commercial approach. But, the more weather-sensitive the operation of the vessel becomes, the more important it is to integrate weather forecasting services into planning.

Once the strategic decision has been taken to follow a given route, it is not normally possible to alter the course radically in the latter stages of the voyage if the weather starts to shift unexpectedly after a few days to make the chosen route much stormier than initially predicted. It may, however, be possible continually to fine-tune the route to obtain the best possible weather within the pre-determined criteria. For example, a container ship with a service speed of 25 knots can usefully refine its route regularly throughout the voyage if necessary. At the other end of the spectrum, there is little which can be done for a tug towing a barge at 4 knots. Once at sea it has to take what comes. The limiting threshold is also a factor which determines whether or not fine-tuning during the voyage is of value. For example, if the limiting wave threshold is a significant height of 8 metres then it is a feasible proposition continually to fine tune the route to avoid this, provided the speed of the vessel is not too low. On the other hand, if the limiting threshold is 2 metres significant wave height this is probably of little use by way of any fine tuning that can be done, even for a vessel with a speed of 25 knots.

Most tows have a speed of 8 knots or less in good weather conditions. Some of them are as slow as 3-4 knots. In heavy weather the speed can drop to near zero. Oceanic tows are usually designed to withstand the motions resulting from the 10-year storm over the route. For coastal tows, or for short tows such as in-field rig moves in the North Sea, the design requirements can be relaxed somewhat. Even when the tow is designed to withstand the 10-year storm every effort is made to ensure that conditions do not approach that value. As a rough rule of thumb, for oceanic tows, the aim is to ensure that the tow encounters nothing worse than a Force 7 wind. When the wind exceeds Force 7 there is then a rapidly increasing probability of the towline parting: the first stage in a sequence of events which often ends up in serious damage or even total loss. So the principal objective must be to ensure the tow does not experience anything worse than a Force 7 wind.

Because of the slow speed of tows it is necessary to take early avoiding action. The forecast for days 3, 4 and 5 can often be more important and more useful than the forecast for days 1 and 2, even allowing for the reducing accuracy in the detail that far ahead. If a tow is likely to experience a severe gale within 24 hours there is not a lot which can be done about it. On the other hand, if the possibility of a severe gale can be identified 4 days in advance there is an opportunity to take some avoiding action. When a tow is in mid-ocean the only avoiding action which can be taken is to alter course. When close to land another option is to seek shelter in a safe haven if one is close enough. If the tow is entirely coastal, no matter how long, it is possible to carry it out with less stringent design requirements provided there is always a safe haven within 48 - 72 hours sailing time. This needs careful advance planning but weather routeing services have gained a great deal of experience in a great many tows which have been successfully carried out in this way. In these circumstances the planning may be dominated by climatology. For example, a tow of an oil rig from, say, Alabama to West Africa in late summer will be defined by the need to avoid any contact with hurricanes (see Section 3.9).   This means taking a route which hugs the coast of the Gulf of Mexico, Central America, and the northern coast of South America, until getting close to the Equator before setting out across the Atlantic. A more direct course would be far too risky, even if the forecasts contained no likelihood of hurricanes within the foreseeable future.

Dry transportations are a different proposition. These are voyages where large items of industrial hardware, such as mobile oil drilling rigs, are carried as deck cargo on specially designed vessels. Typically, the vessels have a service speed of 10 - 13 knots in favourable weather conditions.   This means that it is much easier to avoid adverse weather than is the case with a tow.   Also, although the design requirement, in terms of cargo motions, will be the same as with a tow, the weather threshold which is operationally acceptable will often be higher as there is no towline to break and more severe weather conditions will be required to produce the limiting motions.. This flexibility enables ships' masters to make fuller use of forecasts, both at the tactical and strategic level.   But, if the whole voyage remains plain sailing because the projections beyond five days either prove correct or do not hide difficult conditions, this is fortunate rather than a normal consequence of current forecasting performance.   So, an understanding of the consequences of ensemble forecasting and whether or not the weather is in a predictable form is an important part of making decisions.

This leads into the second aspect of using weather routeing; namely the role of the master.   This remains paramount, but with the development of improved Inmarsat communications and relatively cheap, but powerful on-board computer hardware it is now possible to update shore-based routeing recommendations with the master's analysis of the options available for the ship. So, while many ships' masters may have to rely on an initial forecast and routeing recommendation, increasingly vessels will be equipped with systems which can receive the latest analysis of current and future conditions.   Using these services will involve balancing whatever pressures there are to achieve the scheduled voyage on time, while not jeopardising safety, or damaging the vessel or cargo. In this context, damage includes not only the immediate action of the waves but also the insidious effects of metal fatigue especially in low-dead-weight vessels built with high tensile steels, so the avoidance of slamming may be an important factor in selecting alternative routes.   All of these factors require masters to combine their professional experience with accurate and accessible forecasts of the likely sea-state, wind speed, and direction during any voyage, plus an assessment of how these conditions will affect the progress of their vessel.

Where charterers require the use of weather routeing services the objective is often more simple. What is at issue is whether the vessel has performed to specification. Routeing services are able to propose the optimum route for a given type of vessel and then estimate how long the voyage should have taken given the actual conditions experienced. Given this retrospective analysis will play a central part in any dispute over failure to meet contracted performance, it exerts a substantial influence on sailing decisions. So understanding the limitations in the forecasts and the flexibility open to the ship's master is an essential part of efficient operations.

These differing objectives affect the relationship between forecasters and their customers. Depending on the nature of the vessel, the route, the time of the year, the requirements of the charterer, the interests of the owners, and the confidence of the master in the forecast, the balance between how much emphasis is placed on the forecast of weather conditions and other competing objectives will vary. Where practical, there is much to be said for there being an interactive exchange between the master and the forecasters to explore the limitations of any predictions of heavy weather and the extent to which this may influence thinking on taking evasive action. This process is more likely to work effectively where the routeing service is linked to the forecasting service.

The practical application of forecasts provided by weather routeing services, is considered in Chapter 5. What matters here is that weather-routeing services have to combine not only the skills of forecasting the weather conditions along the route, but also the underlying swell which may have been generated up to ten days earlier by storms on the other side of the world. The fact that these can interact with the wind-generated waves in a complex way (see Section 2.5) to produce locally difficult sea conditions, especially at time of rapid cyclogenesis, requires skilful interpretation by the ship's master. This is an area where there has been considerable progress in recent years. The current generation of numerical wave models (see Section 4.4) now handle swell with considerable accuracy. Nevertheless, it remains a complex subject and, over the open oceans swells will often be encountered which are not predicted in weather forecasts. This frequently occurs when there is swell from more than one major distant weather system present.

4.6. Predicting Storm Surges

Close to shore, especially in parts of the world where there are extensive shallow seas, the combination of tidal forces, variations in surface air pressure and wind stress over the continental shelf cause significant fluctuations in sea level. These have two major consequences for the maritime industry. The first is the need for short term forecasts of extreme surges resulting from stormy conditions coinciding with high tides to enable action to be taken to reduce damage to both installations and vessels. The second is the ability to predict the most likely highest level in, say, 50 or 100 years to provide a basis for making certain decisions about the design of shore-based installations or offshore structures.

How much the sea level rises during the passage of a depression depends on three things: How low the pressure is at the centre of the depression; how fast it is moving; and the depth of the sea. In the right circumstances, usually involving fast-moving systems over waters less than about 100 metres deep, a *resonance* can occur which produces an exceptional rise in sea level above normal levels. When combined with strong onshore winds and high tides, these conditions can cause huge damage. The regions most prone to damaging storm surges are the Bay of Bengal, the Gulf of Mexico and the North Sea. The scale of these surges is evident in Figure 4.6.[8] The disastrous storm of 31 January - 1 February 1953 (Fig. 4.7) killed over 300 people in the UK and about 1800 in Netherlands. This killer storm showed just how vulnerable the coasts around the southern North Sea are to a deep depression moving towards southern Scandinavia and high pressure to the west of Ireland when low atmospheric pressure, strong north to north-westerly winds and the funnelling effect of local topography combine to produce a major surge. The storm surge is clearly seen in Figure 4.6 as an increasingly substantial feature as it swept southwards across the North Sea.

Weather services around the world have developed increasingly sophisticated models to predict accurately damaging storm surges.[9] In the UK the Storm Tide Warning Service (STWS) operates within the National Meteorological Centre of the Meteorological Office. The STWS surge-forecasting techniques involve long established statistical equations for how much the sea level will be raised in any part of the North Sea for a given combination of atmospheric pressure, windspeed and direction. These are combined with the Meteorological Office's standard numerical weather forecasts (see Section 4.2) to estimate the size of the surge and its timing with respect to the standard predictions of high water. The STWS then issues warning of the combination when there is the threat of the predicted surge combining with a high tide overwhelming flood defences.

The large tidal range in the North Sea means the precise timing of the surge, which typically last between six and twelve hours, is an essential part of useful warnings. Where it coincides with a high tide then it constitutes a far greater threat to safety. So, although the numerical forecasts can identify potentially dangerous conditions 36 hours ahead,

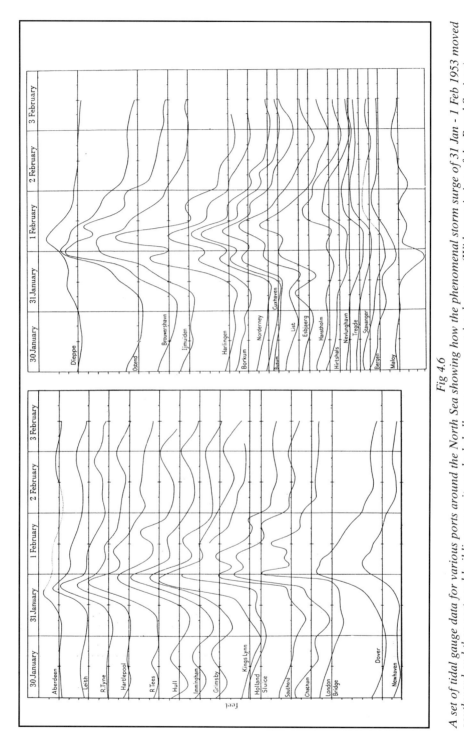

Fig 4.6

A set of tidal gauge data for various ports around the North Sea showing how the phenomenal storm surge of 31 Jan - 1 Feb 1953 moved southward and then eastward building up as it reached shallower more constricted waters. (With permission of the Royal Society.)

the operational work is concentrated in the last 24 hours and depends heavily on the statistical rules. Formal warnings of a flood event have to be issued at least 12 hours in advance and up-dated 4 to 7 hours before high water. The most obvious user of storm surge warnings is the Port of London Authority. They will close the Thames barrier when water levels are predicted to rise above some pre-determined level. This effectively "closes" that part of the Thames to navigation. At other ports abnormally high water levels may make it difficult for vessels to remain alongside some quays. Also, high water levels may allow more swell to penetrate a port than is normally the case, which combined with stormy conditions, increases the risk of damage.

Storm surge warning services exist in other parts of the world. For instance, the National Hurricane Center in Miami includes predictions of surge height in their hurricane warnings.   Because flooding is the

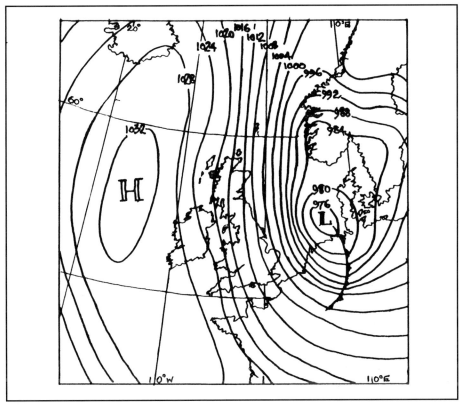

Fig 4.7
The synoptic situation 0000 hours, 1 February 1953 showing how the deep depression over the southern North Sea and high pressure in the North Atlantic creates the conditions for a storm surge in the southern North Sea.

principal cause of mortality when hurricanes make landfall in the USA, evacuation plans rely heavily on forecasts of the expected storm surge to minimise the loss of life. Similar warnings are an essential part of the emergency arrangements in Bangladesh which are designed to enable local volunteers to disseminate cyclone warnings and guide the local populace to constructed mounds or cyclone shelters.

National oceanographic services are also using models which combine extreme weather events and knowledge of the dynamics of a given body of water to estimate the probability of storm surges exceeding given levels. An example of this type of analysis is shown in Figure 4.8. This particular example shows not only how in the North Sea the surge builds up in the south east corner with the area around the mouth of the Elbe being particularly vulnerable, but also the scale of the defences that are needed to protect against a one-in-50-year surge. These defences will, of course, have to be evaluated continually to take account of how the combination of rising sea levels (see Section 6.5), and potential storm surges threaten any particular coastal area.

4.7. Forecasting Currents

Another feature of the interaction between the atmosphere and the oceans that is of immediate interest to maritime activities is the generation of surface currents by transient weather systems. In certain parts of the world these interactions can produce currents which make tidal information in isolation inadequate. These currents can be a major hazard to offshore operations such as tankers, support vessels, and search and rescue operations. So models which can combine the output of numerical weather predictions and the development of ocean currents on the continental shelf are of great value to maritime industries.

An example of this type of forecast has been developed by the UK Meteorological Office in collaboration with the Proudman Oceanographic Laboratory. This uses predicted surface winds and air pressure data to drive the current modelling calculations, together with astronomical tidal information at the deep ocean boundaries. It provides seafarers with a visual presentation of surface currents up to 36 hours ahead. These forecasts provide hourly values of currents and are updated every 12 hours to enable sailing schedules and other activities to be planned more effectively.

Another aspect of the forecasting of atmosphere-ocean interactions is how deep depressions generate considerable sub-surface eddies extending to considerable depths. This phenomenon is of particular interest in the North Atlantic, where these features tend to run along the

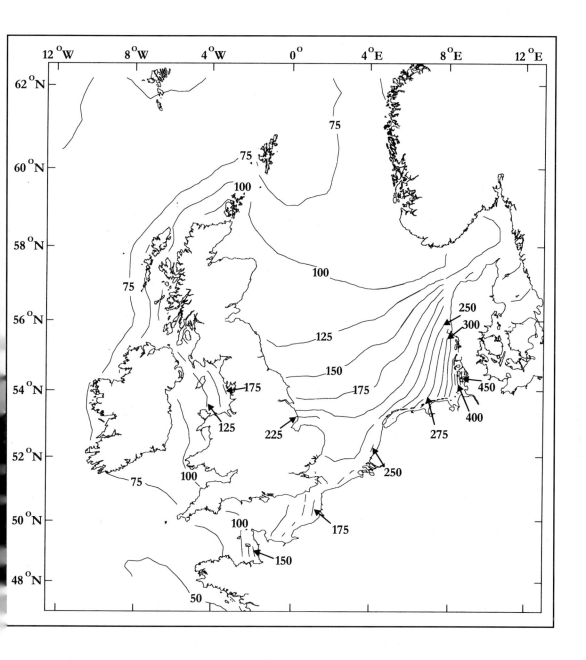

Fig 4.8
An estimate of the increase in sea level (in cm), which could be expected in a 1-in-50-year storm surge in the North Sea. (with permission of the Proudman Laboratory.)

edge of the continental shelf to the west of the British Isles and pose a significant challenge to deep water oil production operations. In the West of Shetland Province where BP are developing the Foinaven field using a tethered tanker system these meteorologically related currents need to be anticipated to enable the system to remain on station. The UK Meteorological Office is developing forecasts of these conditions. It is, however, a measure of the complexity of the processes involved that, while ocean-atmosphere models described above for surface currents work reasonably well for surface currents, they have great difficulty handling conditions along the edge of the continental shelf. For these deep water eddies, the best results have, so far, relied on an empirical approach which uses forecasts just six hours ahead to provide guidance on managing oil-extraction operations. But, as hydrocarbon provinces are developed in deeper waters, the ability to predict these sub-surface phenomena will become increasingly important in the economic exploitation of such resources.

4.8. Seasonal and Longer Term Forecasts

Although, even in favourable circumstances, standard numerical weather forecasts cannot provide any useful guidance more than about a week ahead about future weather, there is a lot of work in progress on longer range predictions. These do not attempt to forecast day-to-day weather but estimate the probability of certain conditions (e.g. temperature, rainfall, incidence of tropical storms etc.) being above or below normal. To do this researchers have examined climatic records for evidence of connections between slowly varying components of the climate, notably sea surface temperatures, and subsequent weather patterns. Many of these efforts have had only limited success, but in the tropics the ENSO is undoubtedly a major factor in seasonal weather patterns around the globe.[10] Most of these relate to the incidence of drought or abundant rainfall, depending on the phase of the ENSO, in places as far apart as Australia, Brazil, India and Zimbabwe.[11] The related question of its influence on the incidence of tropical cyclones and perhaps events at higher latitudes, means that predictions of future ENSO behaviour and hence seasonal forecasting can affect the interests of mariners.

ENSO forecasting is based on both physical model and empirical methods. Building on the physical insights obtained during the events of 1982/83 and 1986/87, modellers have been able to create a variety of detailed simulations of how the atmosphere and ocean interact across the tropical Pacific. Because it is possible to exploit the relevant part of coupled atmosphere-ocean General Circulation Models (see Section 6.4) and empirical data of how past events behave, it has been possible effectively to treat the tropical Pacific in isolation and to tune its response to changing conditions to build up a variety of different modelling schemes. These hybrid systems have proved, with varying degrees of success, that it is possible to predict the occurrence and course of ENSO events.

The performance of the various models is the subject of continual review.[12] What has emerged so far is that different models are all capable of providing useful forecasts. The best results are obtained when strong ENSO episodes occur, including the warming of 1986/87, the cold period in 1988/89 and the rapid warming into 1990. Different models did better in different instances, but they all showed reasonable skill. But when the fluctuations were weaker or the equatorial Pacific quiescent, they were in trouble. In particular, the continued existence of a moderate ENSO warm event from the beginning of 1991 to well into 1995 caught all the forecasters out. More disturbing was the fact that the physical models did not significantly outperform empirically based models. What is not yet clear is whether the ocean-atmosphere system has sufficient inherent predictability to enable physical models to be improved appreciably beyond purely empirical methods. The fact that in 1997 the El Niño returned with a vengeance when all the forecasting systems, apart from the ECMWF, predicted no more than a slow warming in the eastern Pacific shows that the models still have a way to go in encapsulating the true nature of the ENSO phenomenon. Furthermore, the forecasts were slow to pick up the sudden switch to the La Niña conditions in mid-1998, with the ECMWF again being the most successful in predicting this rapid change from January 1998.

One component of these problems is the part played by what are known as intraseasonal variations in the tropical weather. These are pulses of strong winds and rain that travel eastward around the equator in about 30 to 60 days. Known as the Madden-Julian Oscillation (MJO), after the scientists who first identified it, this phenomenon is at its strongest from December to May, when it is one of the most intense weather systems in the tropics, pumping huge amounts of heat into the atmosphere. More relevant is the fact that if one of these bouts of activity, with its strong convection and westerly winds happens to hit the western Pacific just as an ENSO event is ready to hatch, it can stimulate its rapid development. This is what appears to have happened in early 1997. It is, however, in the nature of these MJO oscillations that they are difficult to forecast, being as chaotic as other aspects of atmospheric circulation. So the prediction of ENSO events may still be limited by the inherent uncertainty of modelling the behaviour of the atmosphere.

This unwelcome discovery has, in part, been compensated by the results of the measurements by the TAO array (see Section 3.1). These buoys provided unequivocal evidence of the scale of the warming taking place in the eastern Pacific during the spring of 1997. So, although the ENSO forecasts were in some disarray, the scale of the warming was in little doubt from an early stage in the event, and so the forecasts of its progress could be brought into line by these observations. Indeed, the success of the ECMWF forecast relates in part to the effective use of the data from the TAO array. So, while the long term prediction of ENSO events may be limited by the behaviour of the atmosphere, the inertia of the ocean is such that once it is clear an event is underway, its subsequent evolution is inherently more predictable.

In spite of the problems posed by the atmosphere, the influence of the El Niño means that the development of seasonal predictions will inevitably become part of the meteorological scene. The most obvious potential application of ENSO forecasts to maritime activities is as part of the predictions of hurricane activity in the North Atlantic and Caribbean.[13] Developed by William Gray and colleagues at Colorado State University, these forecasts are built round the fact that there is a strong association between rainfall in the Sahel region of West Africa and the number of intense hurricanes that hit the United States. The hurricane season in the Atlantic starts in earnest at the beginning of August and runs to around the end of October (see Section 3.9). The forecasts use three empirical forecasting routines. The first is produced at the beginning of December for the next year, the second is on 1 June and the final one is on 1 August as the season gets going. Each one uses a different but related set of predictors.

Nine months ahead the forecast is built round the Quasi-Biennial Oscillation (QBO) (see Section 2.5), the rainfall in West Africa in the previous summer and the latest ENSO forecast (warm episodes tend to reduce hurricane activity, while cold events enhance it). The inclusion of the QBO is an interesting addition. This oscillation is not only reasonably regular in its 27 month period, but also the reversal of the winds first takes place high in the stratosphere and propagates downwards over several months. So its effects can be anticipated with considerable accuracy. Furthermore, the fact that the change in direction of stratospheric winds can influence the weather in lower atmosphere by either enhancing or suppressing hurricane activity may be the key to other seasonal forecasts.

By June the forecast is refined to include current ENSO conditions rather than forecasts. It also takes account of pressure patterns and high level winds over the Caribbean and temperature and pressure patterns over North Africa north of the equator. The final forecast is able to use information extending right up to the start of the hurricane season. The most important addition is actual rainfall in the western Sahel in June and July.

All three forecasts had, up until 1997 (see Section 5.6), shown considerable skill in predicting the number and intensity of hurricanes during each season. They did not, however, give any guidance as to when the storms will occur and what paths they will follow: this remains the preserve of the numerical weather forecasts. Nonetheless, the relative success of the December predictions is particularly rewarding, as it suggests that useful warnings of active hurricane seasons can be provided in good time. While the forecasts in June show considerable improvement, the August ones have yet to show significant additional benefit.

The future success of these forecasts depends on an improved understanding on whether atmospheric effects do control the onset of events and the nature of links between the ENSO and the QBO, and also on whether the general behaviour of hurricanes in the tropical Atlantic alters. Gray considers that the decline in activity in recent decades is linked to a weakening of the thermohaline circulation in the Atlantic (see Section 2.2). There is some evidence of the first signs of this process speeding up again (see Section 6.5) and this may lead to increased hurricane activity in the years ahead. So the forecast for 1996 included an analysis of the likely increase of major hurricanes striking the US East Coast, Florida and the Caribbean basin in the years ahead. It concluded that a multi-decadal circulation change could be in progress with the ominous prospect of an increase in these hurricanes. Activity during this season exceeded the level forecast by a wide margin, and Gray proposed that a strengthening of thermohaline circulation in the Atlantic to be a reason for this underestimate. The failure to forecast either the quiet season of 1997 or the active one of 1998 suggests that it is too soon to say whether these longer term changes are yet taking place.

In mid-latitudes the prospects for seasonal forecasts are even less rosy. Although there are identifiable connections between the ENSO and seasonal patterns, notably in the North Pacific, the wider implications are less well established. What is more striking is that the essentially chaotic nature of atmospheric patterns in mid-latitudes appear to dominate the forecasting process. This unwelcome fact has been explored using the Hadley Centre GCM.[14] Using the historic data for global SSTs, going back to 1871, together with more recent sea-ice data, the impact in fluctuations in SSTs on seasonal weather patterns has been calculated. This work showed that these differences were inconsequential compared with the initial atmospheric conditions, which vary substantially from day to day. The ENSO had some impact in winter in the North Pacific and in summer in Europe. But the principal conclusion is that, unlike the Tropics, variations in mid-latitudes between seasonal patterns from year to year are largely the product of the chaotic variations of the atmosphere. The only crumb of comfort is the hint of a signal in decadal patterns. The simulation of winters in the 1960s suggested the prevailing SST anomalies produced increased the tendency for stationary high-pressure systems to form in the vicinity of the Greenwich meridian. This tallies with the North Atlantic Oscillation (NAO) being particularly low during this decade (see Fig. 3.6).

One interpretation of these results is that for most of the effects we are discussing here the driving force for change is fluctuations in the atmosphere. This means that it is shifts in the atmosphere that initiate change, but once they settle down into a given form this generates patterns in the oceans, the cryosphere, and on land surfaces which can either reinforce the new regime or sow the seeds of its destruction. This has two interesting consequences. First, it means that unusual patterns can be sustained for anything from a week or two to many years, but the period is indeterminate. The second is that, in spite of lasting for a long time, the patterns can change suddenly and unexpectedly. The behaviour of, say, the NAO seems to be consistent with this model.

This uncertain picture does not mean that all longer term forecasting is pointless, but it does take us into a world of statistics and probabilities. It means that as with ensemble techniques in shorter-term numerical weather prediction, seasonal forecasting will, at best, provide some indications of whether the coming months or years will feature an above or below average incidence of certain extremes. This type of probabilistic forecasting is, however, only an extension of the types of predictions which are used in many aspects of maritime business. So, beyond the immediate issues of planning routes and avoiding severe weather, the industry may be able to use longer term forecasts to assess risks in, say, reinsurance management. In the longer term this work may influence investment planning, but it may be very long time before this a standard part of the business.

Part of the challenge of using longer term forecasts is defining just what is being predicted and how it might be exploited. For instance, in the case of the insurance industry considering seasonal forecasts of, say, hurricanes, there is the question of interpreting predictions of active or quiet seasons. One single hurricane in a very quiet year, if it were to rip through, say, the Gulf of Mexico oil fields, could cause much more damage than all of the hurricanes combined in an active year if none of them made a direct hit on any areas of substantial insured property.

The same doubts about the using longer term forecasts applies in the case of design of fixed installations, which involve looking 20-50 years into the future. The present technique of carrying out statistical analysis on past records to calculate the magnitude of low probability events corresponding to 50-year or 100-year return periods. This has obvious shortcomings which will be addressed when we consider climate change in Chapter 6. The assumption that the climate of the past few decades is representative of the climate for the forthcoming 20-50 years cannot be justified on the basis of what we know about past changes. But, there is no alternative until we are able to say whether extreme waves in the North Sea, for example, will be higher or lower in 30 years time than they are today. As we will see, for all its limitations, the present system of using past records is probably the best we can do, provided we have records of sufficient quality.

FOOTNOTES.

[1]   Morris (1987).
[2]   Palmer (1994).
[3]   Burroughs (1991).
[4]   Bader et al. (1995).
[5]   Janssen, Hansen & Bidlot (1996).
[6]   Cardonne et al. (1996).
[7]   Wurman & Winslow (1998).
[8]   Rossiter (1954).
[9]   Pratt (1995).
[10] IPCC (1995), p. 215.
[11] IPCC (1995), p. 166.
[12] Barnston (1995).
[13] Landsea et al. (1994)
[14] Mitchell et al. (1996).

MARITIME WEATHER AND CLIMATE

CHAPTER 5

OBSERVING THE WEATHER AND
INTERPRETING FORECASTS

*'Things have their due measure; there are ultimately
fixed limits, beyond which, or short of which,
something must be wrong.'*
Horace 65-8 BC

The objective of this chapter is to show that keeping a close eye on the weather is an essential part of safe and efficient maritime operations. This analysis will focus principally on shipboard activities and how they should interact with weather forecasts. There are, however, good reasons for other parts of the industry to stay abreast of current meteorological developments. These relate to maintaining efficient operations and understanding the threats that adverse weather pose to the maritime industry and its associated financial institutions. So we will also consider the wider implications of maintaining a close watch on the weather. In so doing we will demonstrate that making observations of the weather serves four important and immediate practical purposes:

- it increases the awareness of weather developments, which is of particular value in terms of operating vessels at sea;

- it helps to interpret weather forecasts in terms of local developments and to identify occasions when the weather is evolving in a different manner to that forecast;

- it enables ships' masters to anticipate the need to take emergency action more effectively; and

- if transmitted ashore, it provides a valuable source of measurements which can help weather forecasters in improving their predictions.

There are good operational reasons for developing meteorological expertise. A common factor in many weather-related accidents is that users of available weather forecasts took them at face value and assumed that they would be 100 per cent accurate in weather situations which were marginal for the planned activity. In these cases the accident occurred when the activity went ahead and the weather then deviated adversely from what was predicted. The moral for all users of marine weather forecasts is to assume that no weather forecast will ever be 100 per cent accurate and that there is always a need to gain an appreciation of the limitations of the forecasts and how they should be interpreted when making decisions.

A proper sense of realism comes from a genuine interest in what the weather is doing. Professional mariners will have had some formal training in meteorology and have passed examinations in basic theory and practice. Nevertheless, unless this learning is kept up to date by application to current events, it will become rusty. Whether or not this process occurs depends on how much interest individual mariners have in the weather, which may be largely a matter of personal inclination. So, while recognising that it is hard to change human nature, an underlying objective is to make the whole process of both observing the weather and interpreting the advice supplied by meteorological services more interesting. This objective also extends to convincing ship owners and the insurance industry that it is in their interest to develop within their organisations an increased awareness of how weather hazards can be anticipated and hence avoided in all aspects of shipping operations.

This interest is not just a matter of understanding weather forecasts while at sea. Increasingly, meteorologists and national weather services are exploring the possibility of predicting longer term changes in the weather and the climate. These have important implications for maritime industries in terms of managing risk and making investment decisions.

5.1. Shipboard Observations

There is already ample published guidance on the standard instruments and techniques for making and recording measurements of weather conditions at sea.[1] Here we will not repeat the details of this basic guidance; instead we will look beyond this standard practice at the limitations in the measurements and how this affects their use in improving operations. This involves questions of how to obtain reliable measurements and how to recognise that where there are limits to their accuracy this has a direct bearing on their use in interpreting current events and available weather forecasts.

5.1.1. Pressure

The most fundamental measurement is atmospheric pressure. Accurate and regular measurements provide direct evidence of how conditions are changing in any locality and confirmation of whether or not the weather is behaving as forecast. A standard precision aneroid barometer is fully capable of giving accurate values of sea level pressure, providing the system is calibrated for its height above sea level.[2] It is, however, good practice to make regular checks with standard barometers maintained by Port Meteorological Officers and other harbour and marine offices. This provides an indication of the reliability of the instrument and if there is any evidence of drift this can be allowed for on long voyages. It also addresses the unfortunate fact that inaccurate reports of atmospheric pressure from ships are all too common. This is mainly due to calibration problems.

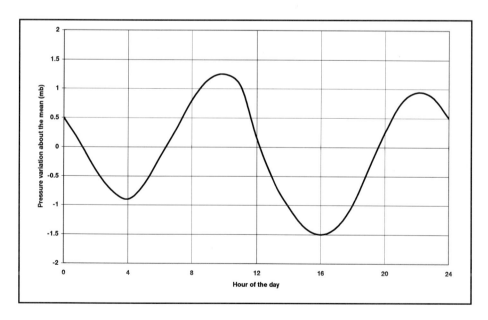

Fig 5.1
A diagram showing the diurnal variation in surface pressure in the tropics.

139

The interpretation of changes in pressure in the context of forecasting will be discussed in Section 5.3. There is, however, one regular variation to consider here. This is the *diurnal variation* in pressure which is superimposed on the irregular fluctuations due to changes in the weather. In mid and high latitudes these variations are usually swamped by changes due to weather systems, but in the tropics they stand out clearly (Fig. 5.1). Their amplitude varies from nearly 3 mb at the Equator to only 0.4 mb at 60°N and S, and their maxima always occur at about 1000 and 2200 hours local time and their minima at 0400 and 1600 hours. The importance of these variations is in the tropics where the pressure remains virtually constant apart from during the passage of tropical cyclones. So any fall of more than 3 mb below the local monthly normal, after correction for the diurnal variation, means there is a real risk that a tropical storm has formed, or is forming in the vicinity (see Section 5.3).

5.1.2. Temperature

Day-to-day changes in temperature are often of relatively little operational interest to most mariners. It can be important, however, when ships are carrying temperature-sensitive cargo in uninsulated holds or containers. Relative and absolute humidity can also matter in some cases, as cargo damage can occur due to excessive condensation. But, in general, only when conditions depart some way from the climatic normal does the temperature become an important factor in decision making. Nevertheless there are a number of features about observing air and sea surface temperatures worthy of mention. The most basic is the challenge of making accurate measurements aboard ship. Although this is rarely a problem for operational purposes it is a major difficulty when seeking evidence of climatic change in the millions of observations made by seamen over the last century and a half (see Section 6.1). So for onboard measurements to be of value to meteorological services, they must be made under standard conditions. This means instruments must be housed in a correct shelter, otherwise observations can be seriously affected by, say, solar heating of the decking. In certain circumstances, these errors can be sufficiently large to be seriously misleading. So it is wise to ensure that, where measurements are being made, adequate efforts are made to standardise the equipment.

The challenge of obtaining reliable observations underlines the value of organised arrangements to make standard measurements on many ships. Notable examples of these schemes include the UK and WMO Voluntary Observing Fleets, which provide benefits for shipping operators of participating countries, as well as to maritime operations more generally. The UK scheme involves nearly 600 vessels, offshore structures and units, while there are 7300 merchant vessels in the WMO scheme. The principal reason for maintaining these observation systems is that the accuracy of weather forecasts is directly linked to the quantity and quality of the observational data from which the initialisation of numerical weather forecasts is derived (see Section 4.1). There is never sufficient good quality observations, so every additional accurate measurement is of potential benefit. Observations from coastal waters are every bit as important as observations from oceanic areas. Forecasters greatly appreciate the observations provided by ships when interpreting the products of computer prediction of the weather. All ships benefit from improved accuracy in the forecasts.

In most instances measurements of temperature at sea serve only to confirm what is evident from other observations of the weather. Where they assume greater importance is when they show the development of a significant differential between the temperature of the air and the sea. In these circumstances, it is the implications for the safety of the vessel because of poor visibility or icing in winter (see Section 3.4 and 3.10) and the safety of the crew that matter. In addition, where the temperature differs appreciably from the climatic normal this is often a sign that the weather is up to no good and hence is a useful check on weather forecasts (see Section 5.3). This is of particular relevance in mid-latitudes in winter where the temperature is well above normal as this indicates a major incursion of air from low latitudes which often combines with adjacent cold air to generate rapidly deepening storms. So maintaining regular temperature measurements and being able to interpret these in terms of standard climatological data is part of good maritime practice.

5.1.3. Wind speed

The majority of wind speeds logged by ships rely on the empirical relationship built into the Beaufort Scale and the related sea-states (Table 5.1). Standard texts on weather at sea (see Bibliography) contain pictures of sea states which correspond to the wind speeds on the scale. The use of these examples is not a precise process. Even with considerable experience of using the descriptions of the state of the sea, it is not always easy to decide which category is applicable, especially when conditions are changing rapidly. Nevertheless, developing this skill is an essential part of good seamanship and hence should be practised whenever at sea. Furthermore, the ability to compare actual wind speeds with the values expected from forecast charts (many forecast charts include a scale for estimating the geostrophic wind in terms of the pressure gradient) provides a useful check as to whether conditions are developing as predicted or taking an unexpected course.

Table 5.1
Beaufort scale: specifications and equivalent speeds

Force	Description	Specification for use at sea*	Equivalent speed at 10 m above sea level				Description in forecasts	State of sea	Probable height of waves* metres
			Mean		Limits				
			knots	metres per second	knots	metres per second			
0	Calm	Sea like a mirror.	0	0.0	<1	0.0–0.2	Calm	Calm	0.0
1	Light air	Ripples with the appearance of scales are formed, but without foam crests.	2	0.8	1–3	0.3–1.5	Light	Calm	0.1 (0.1)
2	Light breeze	Small wavelets, still short but more pronounced. Crests have a glassy appearance and do not break.	5	2.4	4–6	1.6–3.3	Light	Smooth	0.2 (0.3)
3	Gentle breeze	Large wavelets. Crests begin to break. Foam of glassy appearance. Perhaps scattered white horses.	9	4.3	7–10	3.4–5.4	Light	Smooth	0.6 (1.0)
4	Moderate breeze	Small waves, becoming longer, fairly frequent white horses.	13	6.7	11–16	5.5–7.9	Moderate	Slight	1.0 (1.5)
5	Fresh breeze	Moderate waves, taking a more pronounced long form; many white horses are formed. Chance of some spray	19	9.3	17–21	8.0–10.7	Fresh	Moderate	2.0 (2.5)
6	Strong breeze	Large waves begin to form; the white foam crests are more extensive everywhere. Probably some spray.	24	12.3	22–27	10.8–13.8	Strong	Rough	3.0 (4.0)
7	Near gale	Sea heaps up and white foam from breaking waves begins to be blown in streaks along the direction of the wind.	30	15.5	28–33	13.9–17.1	Strong	Very rough	4.0 (5.5)
8	Gale	Moderately high waves of greater length; edges of crests begin to break into spindrift. The foam is blown in well-marked streaks along the direction of the wind.	37	18.9	34–40	17.2–20.7	Gale	High	5.5 (7.5)
9	Strong gale	High waves. Dense streaks of foam along the direction of the wind. Crests of waves begin to topple, tumble and roll over. Spray may affect visibility.	44	22.6	41–47	20.8–24.4	Severe gale	Very high	7.0 (10.0)
10	Storm	Very high waves with long overhanging crests. The resulting foam, in great patches, is blown in dense white streaks along the direction of the wind. On the whole the surface of the sea takes a white appearance. The 'tumbling' of the sea becomes heavy and shock-like. Visibility affected.	52	26.4	48–55	24.5–28.4	Storm	Very high	9.0 (12.5)
11	Violent storm	Exceptionally high waves (small and medium-sized ships might be for a time lost to view behind the waves). The sea is completely covered with long white patches of foam lying along the direction of the wind. Everywhere the edges of the wave crests are blown into froth. Visibility affected.	60	30.5	56–63	28.5–32.6	Violent storm	Phenomenal	11.5 (16.0)
12	Hurricane	The air is filled with foam and spray. Sea completely white with driving spray; visibility very seriously affected.	—	—	>64 and over	32.7 and over	Hurricane force	Phenomenal	14.0 (—)

* These columns are a guide to show roughly what may be expected in the open sea, remote from land. Figures in brackets indicate the probable maximum height of waves. In enclosed waters, or when near land with an offshore wind, wave heights will be smaller and the waves steeper.

Statistical studies of wind speed estimates using the Beaufort Scale specifications against physical observations where they are available suggest that the average scatter is a little over half the range for any given force.[3] This means that at Force 1 or 2 the accuracy is unlikely to be better than ± 1 knot - an error of 20 to 50 per cent - but this inaccuracy is of little practical consequence. By the time we reach Force 5 to 7 the accuracy will be about ± 3 knots - an error of 10 to 16 per cent either way. For higher wind speeds (up to Force 10) the error remains around plus or minus 10 per cent.

Anemometers in normal operating conditions do not provide additional accuracy. Because they are often fixed at non-standard heights above the water on structures which may have a significant aerodynamic impact on the readings they should be viewed with caution. As a general rule they may do no better than ± 4 knots, at any wind speed above a few knots. Even in scientific studies, where great care is taken with the siting of fully calibrated equipment, aerodynamic effects caused by the ship's structure and motion can have an appreciable impact.

5.1.4. Wave height, period and direction

The shipboard observations of wave height are almost always made by eye. This is no easy matter, especially in heavy seas. Measurements are best performed amidships where the effect of the pitch and roll of the ship can be minimised. Then, by knowing how far above the sea you are, it is possible to estimate the height of the more significant waves (see Section 2.5). This is not an exact science. Studies of observers' estimates against measurements by other physical methods suggest that visual estimates tend to underestimate low wave heights, while big waves tend to be overestimated. The crossover point is about 8 metres (26 feet), which means the great majority of visual estimates of wave heights are too low.

In principle, the period of these waves can be measured by the simple expedient of timing the passage of a given number of them past the ship. In practice, the motion of the ship complicates the situation tremendously. Because the speed of travel of the waves in knots is three times the period in seconds (see Section 2.5) waves in the 4-7 second period range are travelling at speeds similar to that of many ships. As a consequence, the periods reported by ships have been found to be shorter than the actual periods. Furthermore, the range of error is substantial.

Similar difficulties are encountered with estimates of the direction of the waves. Normally, it is that with wind-generated waves it is equated to the wind direction. But, even in this case, differences of up to 20° in the direction of their peak energy with respect to the local wind direction can be encountered. As for the underlying swell this can come from almost any direction depending on where it has been generated by distant major weather systems.

5.2. Single Site Forecasting

Before considering how shipboard observations can be combined with available forecasts, it is useful to remember how much can be achieved without any external sources of information. With a good knowledge of the relevant features of the weather it is possible to make short-term forecasts of local conditions. By observing changes in the sky, the wind and the pressure, a meteorologically aware mariner can make a decent stab at probable weather trends a few hours ahead.

Signs of quiet weather are associated with stable air masses (i.e. thermodynamically warm - see Table 2.2) and a steady or slowly rising barometer. In addition the mariner should look for:

- when in sight of land, the amount of clouds decreases during day, and the clouds rise on mountains;

- near normal temperatures;

- light winds;

- clear skies and an absence of high cirrus which forms a halo round the Moon at night or the Sun by day; and

- in mid-latitudes, clear skies in evening with red sky at sunset, although in mobile westerly weather patterns the duration of the quiet spell may be only a few hours as the axis of a brief ridge passes by.

Signs of bad weather are usually associated with unstable air masses (i.e. thermodynamically cold - see Table 2.2) with considerable activity and a rapidly falling barometer. In addition, the mariner should look for:

- rapidly moving clouds, with the clouds at different levels moving in different directions;

- clouds lowering and thickening;

- temperatures far below normal at any time of year.

5.3. Interpreting Weather Forecasts

As has already been made abundantly clear the interpretation and application of weather forecasts is not a matter of trying to outsmart the forecasters. The product of numerical weather forecasting operations contains the comprehensive analysis of a huge amount of data (see Section 4.1). But, in spite of all this effort, the output has obvious limitations. Furthermore, only a tiny part of the output can be made available to mariners, who will also have widely different levels of information available to them. Here we will assume that most readers are equipped with at least standard fax equipment. This means they will be able to receive mean sea level forecasts for the region where they are sailing. In addition, depending on the locality, they may be able to receive upper air charts, satellite pictures, and a range of material for specialist users. As will be made clear, however, much of this information must be used with great care. So, what matters is asking the right questions about what is contained in the forecasts and how this influences our thinking. This process is not just a matter of trying to identify weaknesses in any prediction but also of extracting the maximum out of the product in making effective operational decisions.

The first thing to get straight is whether we are dealing with a general forecast for an extensive area for no specific activity, or a forecast for a given location and a specific activity. In the case of the latter, where the user will often have direct contact with the forecasters and can seek advice on the uncertainties inherent in the prediction, the approach is entirely different. The ability to explore with the forecasters various aspects of their predictions and the probabilities of them being correct. With broadscale forecasts prepared for the 'mass market' this is not easily done. Here we will concentrate on reading the general forecasts, but the issues addressed are designed to provide a checklist for mariners using specific forecasts for particular activities.

The second factor is how close to shore the mariner is operating. On the open seas it is much easier to interpret the forecasts and it is safer to assume that the direction and strength of wind and waves will be approximately as predicted. But closer to shore local topography can alter the winds and shallower waters and tidal flows can radically modify wave characteristics. Extreme situations can involve winds, strong tidal flows and underlying swell all travelling in different directions creating what can only be described as a 'confused' sea state. In these circumstances, where difficult or risky operations are to be conducted, the use of specific forecasts makes far more sense and close consultations with forecasters is a sensible precaution.

Thirdly, there is the delay between when the forecast is prepared and when it is used. Typically, there is a six hour gap between the time of the base data and the issue of the forecast, and this delay grows until the next forecast is prepared. When the weather is changing rapidly, the mariner will have real observations of the intervening period which can provide invaluable insights as to how accurate the forecasts are likely to be. In this analysis, it is important that the users of weather forecasts not only understand the content of each forecast but also understand the broadscale meteorological processes at work behind it. With this knowledge it does become possible to fine-tune the forecast by careful observation of the local weather. If the user has access to charts received by radio-facsimile this task is much easier but even without charts it is still possible. Some of these broadcasts include satellite imagery, which sometimes contain more up-to-date information than is being used for the latest published forecast. Although not of particularly high resolution this satellite imagery can sometimes give a useful indication of the position of fronts and low pressure centres.

The dividing line between marine activities which are adequately served by general area forecasts and those which require a site-specific forecast service tailored to the requirements of the project is not sharply defined. The general area forecasts are free whereas site-specific forecasts are generally provided on a commercial basis. To justify the cost of the site-specific service there has to be an identifiable benefit gained from it.

One difference between general area forecasts and site-specific forecasts is that the former typically cover only a period of 24 or 48 hours from the time of issue whereas the latter can be commissioned to cover up to 5 days ahead, or even longer in general terms. While it is possible to get general area forecasts five or more days ahead, these are not normally part of standard services to mariners. This difference is important in the context of weather sensitive operations lasting several days, such as oil platform emplacement or for low-speed tows.

General area forecasts are appropriate for any marine activities which are significantly affected only by very severe weather, and which require only a relatively short warning time of its likely occurrence. The routine operation of conventional merchant ships comes into this category.

Site-specific forecasts are appropriate for any marine activities which have one or more of the following characteristics:

- relatively low limiting weather thresholds (e.g. wave height 3 metres)

- duration of weather-sensitive activity greater than about 36 hours

- a tow with a passage speed in fair weather of less than about 8 knots.

These requirements relate not only to voyages but also to weather-sensitive port operations (e.g. loading dangerous cargoes, and pilot work). Site-specific forecasts, which provide warning of approaching sudden severe weather (e.g. squall lines), can be a vital part of safe operations. The absence of such warnings can pose significant risks for charterers. In a recent case the lack of site-specific forecasts was deemed to be a primary contributory factor to the loss of a vessel while discharging gasoline at an exposed port. The arrival of a sudden severe squall caused the moorings to break. The discharge hose ruptured and the gasoline ignited resulting in a major fire in the accommodation section of the vessel. The tribunal concluded that the absence of a local forecasting service able to provide warnings of sudden deterioration of the weather made such an exposed docking facility unsafe. For this reason the charterers were found to be in breach of the charter agreement by ordering the vessel to unload at an unsafe port.

Against this background, whatever form of weather forecasts you are relying on, it follows that keeping a close observation on the evolving local conditions is the best guide to whether or not the weather is behaving according to the forecast. Points to look for:

- is the wind backing or veering more or less than forecast?

- is the wind speed increasing or decreasing faster or slower than forecast?

- is there a swell present which is inconsistent with the predicted weather evolution?

- is the sky changing in a way which is different to what might be expected from the forecast?

- is the pressure falling or rising faster or slower than expected from the forecast?

- has rain stopped or started sooner or later than forecast?

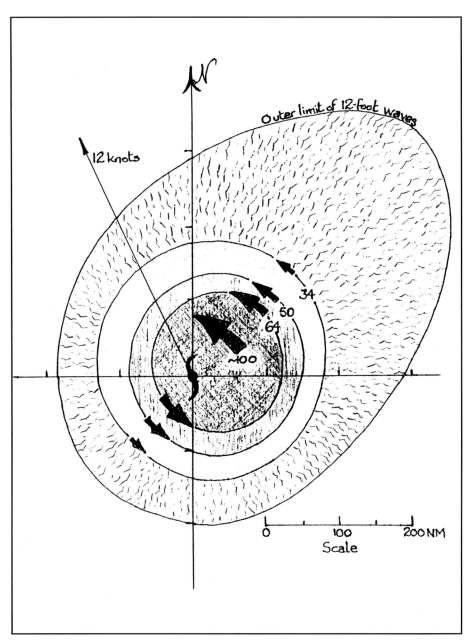

Fig 5.2
A schematic diagram showing the extent of the winds and waves around a category 3 hurricane in the northern hemisphere. The central pressure of the storm is 965mb and it is moving on a course of 330° at a speed of 12 knots. The greatest area of high winds and waves are in the northeast sector (i.e. to the right of the storm track), with the region of highest sustained winds of about 100 knots being found in this quadrant.

Any or all of these can give an indication of how the pressure systems are developing relative to how they were predicted to evolve. Furthermore, where these rules are applied to rapidly developing low pressure systems, they provide the basis for deciding whether evasive action is wise. This applies most obviously to forecasts of tropical cyclones, where the need to steer clear of the strongest winds is vital, and forecasting their course is still a very inexact science. As noted in Section 4.2 the average errors in tropical cyclone position predictions are about 300 nautical miles (550 km) for 72 hour predictions. So, given the scale of these uncertainties, it may be better to combine the rules above with the simple model of how the winds around these storms are normally distributed (Fig 5.2) when deciding on the best course to follow to avoid the worst of the storm.

5.4. Out of Sight, Out of Mind

The guidelines set out above are all well and good in principle, but it helps to have some practical examples. In particular, there are cases from the past which provide insights into why forecasts are not up to scratch and underline what mariners should be looking for. Common to many of these examples is the fact that for one reason or another the forecasters do not have adequate information about what precisely is happening, where storms may be brewing up rapidly. Moreover, for mariners in the vicinity of such developments, their observations are much more relevant to their particular predicament.

A historical example provides a good insight into what we are talking about. In September 1938 a massive hurricane devastated much of New England. Forecasters completely failed to provide any warning because, although the storm had been reported on for several days, by the time it was tracking northwards off the eastern seaboard of the USA, all shipping was giving it a wide berth. So when it continued northwards instead of swinging eastwards as would normally be the case, there were no reports of its progress and hence no way of knowing until it hit an unsuspecting Long Island and New England with catastrophic consequences. In Providence, Rhode Island, the storm surge crested at 4.2 metres (nearly 14 feet) above mean highwater, and the downtown area was flooded to a depth of 3 metres (10 feet). Across New England some 250 million trees were blown down.

One other historical example of the difficulties associated with not having a clear idea of the whereabouts of a tropical cyclone is probably worth recounting. This is Admiral Halsey's disastrous misreading of Typhoon *Cobra* in December 1944. The armada under his command on the battleship *New Jersey* received warning of the developing storm, but misinterpreted the wind patterns. This was because the system overtook a weak cold front and as it intensified it produced confusing winds ahead of it. As a result the fleet, in taking what was thought to be evasive action, steered directly into the path of the typhoon. Three frigates sank with the loss of over 800 men, and some 150 planes had to be ditched from the escort carriers (converted freighters). The damage to the task force was so great that it was unable to participate in the invasion of the Philippines. There was one benefit of this disaster, however, in that the US Navy decided thereafter to maintain aircraft reconnaissance flights to monitor the position and movement of tropical cyclones.

Satellites and aircraft now ensure that forecasters are not operating in the dark, but tropical cyclones still represent a major challenge given the erratic courses they sometimes follow and the relatively narrow swathe of severe damage they cause. Two examples come to mind. The first was typhoon *Vera* in 1986 . *Vera* was unusual in a number of ways and proved to be a major forecasting challenge. It formed in the wake of Typhoon *Tip*, and initially formed a rather ill-defined system between 15 and 17 August, which nearly disappeared, before undergoing a resurgence over the next three days. To begin with it moved eastwards until 22 August, reaching its maximum intensity of 110-knot winds and a mean sea level pressure of 923 mb on 21 August. Later, after more normal WNW movement it recurved more sharply than expected over the East China Sea, crossing the island of Okinawa late on 25 August. It then accelerated and moved north east across the north of South Korea. A belt of Force 9-10 S-SSW winds unusually far from the centre caused extensive damage to shipping in the port of Ulsan. The problems associated with forecasting this typhoon were attributed to both difficulties in identifying the formation of the storm on satellite images (for several days it was not possible to see real evidence of circulation), and then the unexpectedly strong influence of an intense upper atmosphere monsoon trough extending throughout the entire western North Pacific for most of August.

The second example was typhoon *Gay* in 1989. Unique because of its small size, intensity, and point of origin, *Gay* challenged forecasters by crossing two different tropical cyclone basins and almost entering a third. It formed in the Gulf of Thailand on 1 November and became the worst tropical cyclone to affect the Malay Peninsula in 35 years. From a climatological point of view, an occasional tropical cyclone may move into the Gulf of Thailand from the South China Sea, but it is rare for genesis and intensification to occur in the Gulf - a relatively small body of water surrounded by land on three sides.

Less than 24 hours after formation *Gay* had intensified to a typhoon. During this process it presented a paradox to forecasters. While the satellite intensity estimates correctly diagnosed intensification, the synoptic data in Malaysia and Thailand indicated weakening winds and higher pressures. The synoptic data were correctly interpreted as indicators of increased subsidence produced by the intensifying storm system. The cyclone made a direct hit on the drillship *Seacrest*. The ship sank in confused seas with estimated wave heights of 11 to 14 metres (35 to 40 feet) with only 3 or 4 survivors. *Gay* then crossed the Isthmus of Kra with wind speeds of over 100 knots, causing heavy loss of life. Although the passage over land weakened the storm, it began to intensify steadily as it crossed the Bay of Bengal. It achieved super typhoon intensity in the morning of 8 November, and the central pressure fell to 898 mb just before it struck the sparsely populated Indian coast some 200 km north of Madras with sustained winds of about 140 knots. It then maintained its identity as it crossed India, and was forecast to re-intensify off Bombay but, instead, lost its identity just before making it to the coast.

The message that emerges from these examples is that any shipping unfortunate enough to be in the vicinity of a rapidly evolving tropical cyclone should make fullest use of the guidelines set out in Section 5.3, both to identify how close the system is and which is the best course to steer to avoid damaging weather. Even today, with satellite and aircraft coverage, it is possible for tropical cyclones to catch forecasters out. As we have seen these problems can arise for a variety of reasons, including lack of reliable surface measurements, difficulty in interpreting satellite images, and failure to reflect accurately the behaviour of the upper atmosphere.

These problems can also occur when disturbances move to higher latitudes. There is sometimes a gap in the forecasting process between them being a decaying tropical storm, and those relatively rare, unpredictable occasions when they regenerate as powerful extratropical depressions. This is a complicated process involving the plume of very moist, warm air extending high in the atmosphere, and being drawn to high latitudes by tropical cyclones. Because this may transfer energy to adjacent weather features in a way which throws the forecasts off the scent. Hurricane *Lili* in October 1996 (see Section 2.6) is a good example of this phenomenon. Meteorologists have spent a lot of time trying to work out why the models failed to predict why *Lili* did not decay but underwent an explosive rebirth, without coming to any simple explanation. Moreover, there is a sense of there being an air of detachment about events in the middle of the North Atlantic which pose no immediate threat to populous areas of North America or Europe. At a more practical level any shipping in the vicinity of such a transition is far better equipped to see how the forecast is failing to anticipate developments. So simple measurements of pressure together with wind speed and direction combined with correct interpretation of what the weather is up to and how it is diverging from the forecast, is the best guide in these circumstances. What is more, if any such observations are transmitted ashore, whenever it is identified that the actual weather is diverging from the forecast weather, this is of great potential value to forecasters and hence to other mariners.

The challenge of using onboard measurements to anticipate sudden changes far from land also applies to another difficult area for forecasters the rapid development of secondary lows (see Section 2.4). This is particularly true for cyclogenesis in the central North Atlantic and Pacific Oceans. By comparison the same process in the Gulf of Mexico and off the eastern seaboard of the USA, where there are more surface observations, is less likely to catch the forecasters out. A good example of this type of development is a depression which hit the south-west approaches of the British Isles on 13 and 14 August 1979. Because this struck the Fastnet Ocean Race involving over 300 yachts and some 3000 crewmen it is known as the 'Fastnet Storm'. By the time it had blown out 15 sailors had been killed, 24 boats sunk or abandoned and, of the 303 boats that set out from Cowes, only 85 finished the race at Plymouth. One or more crew was washed overboard from 51 yachts and 136 had to be rescued by helicopter or ship.[4]

The storm had its genesis as a shallow secondary feature that developed west of Newfoundland on 12 August and ran quickly across the Atlantic without deepening appreciably, but as it approached south-west Ireland on the evening of 13 August, it developed explosively in a region virtually devoid of observations. By midnight it brought Force 11 winds to the Celtic Sea, and a number of meteorological stations in the region observed their highest August wind speeds on record. By this time the competitors in the Fastnet Race were strung out from Cornwall to approaching the Fastnet Rock and many of them caught the full brunt of the storm.

The weather forecasts at the time did a reasonable job in predicting the path across the Atlantic, but completely failed to provide lengthy warning of the explosive development as it approached south-west Ireland. This is hardly surprising as storms of this intensity are exceedingly rare in this part of the world in August. Moreover, in 1979 the computer models were considerably less sophisticated than now. As a consequence, the first gale warning was not broadcast until 1505 hrs on

13 August after the news on BBC radio and was included in the shipping forecast at 1750. By this time crews in the race had been observing plummeting barometric pressure for several hours. The first Force 9 warning was issued by the UK Meteorological Office at 1805, broadcast after the news on Radio 4 at 1830, and by the time it was included in the shipping forecast at 0015 on 14 August, the warnings had been increased to locally Storm Force 10 in areas Sole, Fastnet and Shannon. But, as Sir Edward Heath the captain of the British Admiral's Cup team, on board the Morning Cloud noted, within four hours, they were encountering gusts of well over 60 knots.

The reason for spelling out these events in detail is to highlight how the combination of limited forecasts, exceptionally severe weather and

Fig 5.3
A satellite image of a polar low in the Barents Sea on 27 February 1987 (with permission of the University of Dundee.)

boats already committed to a given course can lead to tragic consequences. In these circumstances there were only two options open to many of the competitors. The first was to ride out the storm as best they could. The second was, where there is still some possibility of taking evasive action, to use on board measurements to interpret the latest warnings to identify the most effective way of steering away from the most dangerous weather. In the Fastnet Storm those yachts not caught in the immediate path of the strongest winds scattered in all directions seeking shelter in the harbours and coves of southern Ireland, Cornwall and West Wales.

Another example of dangerous systems which can develop rapidly, effectively out of sight and with little warning, are polar lows. As explained in Section 3.4 these systems can measure from around a hundred to several hundred kilometres across, but even in their smallest form can exhibit remarkably organised circulation with a central 'eye' (Fig. 5.3) and winds up to 60 knots, even though the pressure drop across the system may only be as little as 5 mb. These conditions are combined with driving snow and zero visibility which makes them exceedingly hazardous to shipping. But because their size is little more than the grid spacing of the global numerical weather forecasts (see Section 4.1) they are exceedingly difficult to predict accurately. The emergence of more detailed limited area models for shorter term forecasts may be able to get to grips with these systems, but the fact that there is often a complete absence of observational data from within the circulation of these systems means progress is likely to depend on better measurements rather than better models. So, yet again, these weather systems represent another example where there is no real substitute for maintaining a close eye on conditions when sailing in higher latitudes, especially when the air temperature is way below the sea temperature and there is vigorous convection in the vicinity.

Polar lows are not just a hazard of polar waters, but can develop in strong flows of arctic air moving towards low latitudes. In particular, these conditions occur down into the North Sea and the Baltic, especially in late winter and early spring. A good example of this type of system developed in late March 1985.[5] During 28 March what looked like an innocuous collection of convective clouds moved swiftly from south of Iceland to west of Denmark, without giving any warning of its impending transition. Then within six hours it had developed into a vicious little depression which swept across Denmark and into the Baltic. As it crossed Copenhagen, the barometric record (Fig. 5.4) looked just like a tropical cyclone, although the pressure drop was small, there were gale-force winds, thunder and lightning, heavy snow and near-zero visibility.

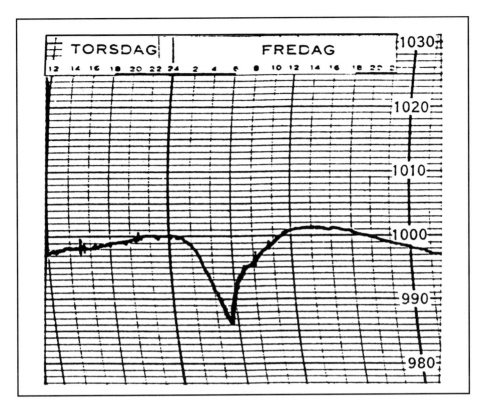

Fig 5.4
The barograph trace for Kastrup Airport, Copenhagen showing the passage of the polar low on 29 March 1985 (with permission of the Royal Meteorological Society).

The fact that this disturbance was not forecast and moved so fast shows that it is not possible to take action to avoid such weather. So the obvious message is there is no alternative to maintaining a continual watch on the weather whenever it is capable of generating sudden storms, even in areas as sheltered as the Skaggerak.

The other place where vigorous small-scale depressions can form suddenly is in the Mediterranean. These occur most often in the western basin around the Gulf of Genoa, but can occur farther east. As noted in Section 3.5 they have been likened to either tropical cyclones or polar lows. They feature the combination of vigorous convective activity, organised circulation, but relatively a small drop in central pressure. This means they are likely to be missed in the numerical weather forecasts, and hence are devilishly difficult to forecast, and can only be anticipated by constant vigilance.

5.5. Medium Range Forecasts

The progress in numerical weather prediction (Section 4.1) has had its most notable benefits in terms of improving forecasts two to six days ahead. Here, the consequences of these improvements are most apparent in terms of routeing. So it is in terms of these services and the underlying products of the national weather services that we will consider the interpretation of medium range forecasts.

The essential feature of many of these forecasts is whether the circulation patterns are in a mobile or static frame of mind (see Section 4.2). This means that what matters is whether the forecast broadly represents the tracks of future depressions together with the timing of their genesis and development, and the position and movement of the main anticyclones. It matters less whether the forecast position of any given system is delayed or advanced by several hours, providing that the broad patterns remain as forecast for five or more days. By comparison a forecast which fails to predict a significant shift of tracks followed by the lows while predicting the timing of their birth and life-cycle accurately is of less value. This is because on a long winter voyage, say, from San Francisco to Yokohama, having made the decision to follow a southerly route to avoid a set of lows which are predicted to move in sequence into the Gulf of Alaska and then down the west coast of Canada, a shift in the steering patterns to carry them on a more southerly course and then turn northwards to Canada would have serious implications.

The first signs of medium range forecasts shifting their view four or more days ahead often shows up in significant changes in the predictions of the circulation patterns at the 500 mb pressure level. So the ability to interpret these standard forecast products and appreciate how changes in successive forecasts can provide the first clues of a shift in weather regimes that could have significant consequences for voyage times is important. While these insights may come too late to enable ships' masters to take evasive action once committed to a given route, they can provide ship operators and charterers with advance warning of delays or even damage to urgent cargoes.

This ability to understand the limitations in forecasts becomes of greater value in making decisions about one-off operations, such as oceanic tows, where the slow speeds mean that the 3 to 5-day forecast is more important than shorter term predictions (see Section 4.6). Furthermore, given the reliance on specific forecasts for planning voyages, there is good reason for operators to have staff who can interpret these predictions and discuss their limitations with the forecasters. This cannot go as far as operators having staff trained sufficiently that they can interpret in-depth medium range forecasts and the science behind them. Instead, operators need to understand that medium range forecasting is an inexact science and that precise predictions are impossible. What is essential is that they have close contact with their forecast providers ensuring that they specifically address the probability of any significant weather thresholds being exceeded. It is up to meteorologists to consider the implications of the 500 mb flow (see Sections 3.2 and 4.2), not the operator. But, in considering the uncertainties in any prediction, it would be appropriate for the forecaster to discuss the 500 mb flow when talking to the operator. In these circumstances, it is helpful to know something about what is being described, although it is asking a bit much for the operator to employ staff trained fully to understand the underlying science.

5.6. Using Seasonal Forecasts

The science of seasonal forecasting is still at an early stage of development. As such it is questionable as to whether it has reached the point of being of practical use to the marine community. It may be some considerable time before it is of any real value, if ever. Nevertheless, there is considerable work in progress and this will, from time to time, receive widespread publicity. In particular, the occurrence of events associated with the ENSO will be the subject of continued speculation. Because these forecasts touch on meteorological factors which are of direct interest to maritime industries (e.g. tropical cyclones), we need to consider whether it offers possible benefits for future maritime operations. So we will review the current state of progress in this area, concentrating on the specific example of the implications of forecast performance associated with the 1997 ENSO event.

As explained in Section 4.7, most of the work on seasonal forecasts relies in identifying statistically significant links between longer term weather patterns and slowly varying components of the global climate system. So, to understand the potential of development work on these forecasts, we need to expand on the reasoning behind their preparation presented earlier. In the light of this analysis it may then be possible to assess progress and decide whether they will have any value to maritime industries. This exercise is not something for mariners at sea to worry about. Planning departments of shipping companies or the insurance industry may, however, need to do some work because, whatever the limitations of this forecasting work, one thing is certain - they are bound to be the subject of intense discussion and publicity in the future.

The essence of interpreting seasonal predictions is understanding how the various factors *(predictors)* are used to prepare a forecast. The number of predictors used, the contribution each makes to the forecast and how reliable each one is in linking current conditions to future weather patterns, are all part of this analysis. This is rarely an easy task as the links are complex and forecasters are inclined to use a variety of overlapping predictions to reinforce their efforts. For this reason, it is best to consider an example of these forecasts. Here we will consider William Gray's work on hurricanes in the Atlantic Basin, which was described in Section 4.7, because it provides a good example of the analytical challenge and also is of direct relevance to several maritime industries.

The Gray forecasts use a pool of 17 predictors to forecast 9 parameters including the number of tropical storms, hurricanes and intense hurricanes plus the number of days with each of these categories together with two measures of destruction potential and one of overall net tropical cyclone activity. Many of the predictors and the forecast parameters are closely related and so only some of the predictors are used to determine specific forecast parameters. Typically the number is five or six and when the observed values of the predictors for each year from 1950 to 1995 are used to make retrospective forecasts *(hindcasts)* this selection can explain between 53 and 68 per cent of the variance in the various measures of hurricane activity in these years. The most widely used predictors are the QBO, the rainfall anomaly in the Gulf of Guinea during August to October of the preceding year, and sea surface temperature anomalies in the mid-Atlantic and mid-equatorial Pacific. Forecasts are issued in December,[6] some eight months before the hurricane season and then updated at the beginning of April, June and August. They provide comprehensive details of the latest figures for the predictors together with detailed argumentation of how this evidence has been used to prepare the forecast for the coming season. So anyone who has an interest in hurricane activity in the Atlantic Basin can keep track of how the forecasters' thinking is developing and, if they so wish, form their own opinion about how the season may develop as the predictors change in the preceding months.

In this context the forecast of a third successive above-average activity season in 1997 provides an interesting example of how slow progress is likely to be. Gray's forecasts remained rock steady up until June 1997, in spite of clear signs of the rapidly emerging major El Niño. Even at the beginning of August, Gray remained convinced that the warming in the Pacific would not cancel the effect of the other predictors, and the season would feature above normal activity. In practice, after initially starting off quite busily with four relatively minor storms in June and July, the season came to an almost complete halt with just a single reasonably intense hurricane in mid-September *(Erika)*. The fact that the switch to La Niña conditions in 1998 seems to have produced an unexpectedly active season suggests that these forecasts require a more fundamental reappraisal.

Gray now considers that while the suddenness of the 1997 and 1998 events overwhelmed his forecasting method, in more normal circumstances, the technique is capable of producing useful predictions. But the ability to anticipate rapid changes in the Pacific in good time is clearly a dominant factor. This suggests that the ENSO forecasts are the key to progress (see Section 4.7). The genesis of the 1997 event does, however, raise serious doubts about whether long-term useful forecasts are a realistic prospect. As explained in Chapter 4 the crux of the issue is what triggers the sudden onset of an event. If atmospheric activity in the western Pacific in the spring of 1997 launched the event by sending a particularly large pulse of warm water moving eastwards then longer term forecasting is under threat. Modelling the combination of atmosphere and ocean in the Pacific up to a year ahead ceases to be a realistic prospect. Instead, if events are triggered by weather developments that cannot be forecast more than a few days ahead, seasonal forecasts will have to rely on insitu measurements of the ocean which show categorically whether or not an El Niño has started. This may mean that, while an ENSO event, once set in train, may follow a well established pattern with predictable consequences, forecasting its onset months ahead will remain unattainable.

This raises fundamental questions about the historical view that the oceans could be regarded as a slowly moving system, which acted as a giant flywheel in steadying the climate. This underlying assumption that the behaviour of the oceans is in some way more 'predictable' has come under increasing scrutiny recently. If the oceans undergo significant changes on as short timescales as the atmosphere and have to be modelled with the same amount of detail, then the prospects for longer term forecasts may be even more remote than the guarded analysis above suggests. Large scale oceanographic studies do show that the oceans exhibit turbulent motions on every scale, on all timescales, and at all depths. This means the behaviour of the oceans may effectively be as chaotic as the atmosphere, even though events like the 1997 ENSO have so much inertia that, once set in train, they are bound to influence the weather for many months ahead.

All this implies that any progress on predictions about the

probability of certain types of seasonal weather, based on the more predictable behaviour of the oceans, will make erratic progress. But this will not stop the forecasts being made and being given wide publicity. A recent example of work published by a group at the ECMWF [7] provides a useful insight into this process. This paper, which received considerable publicity, showed that, in building on the successful forecast of the developing of the warming event (see Section 4.8), the Centre's seasonal forecast produced accurate forecasts of above and below normal rainfall around the world six months ahead. These results, together with accurate US forecasts of the winter of 1997/98 across North America, may, however, be more a matter of the dominant influence of the record-breaking ENSO event during 1997. The real test of all these seasonal forecasting activities will come when there is nothing dramatic happening in the equatorial Pacific. If they can then produce useful forecasts up to a year ahead then this will indeed represent real progress.

In the meantime, we will have to exercise careful judgement about the significance of any published forecasts. The fact that they may predict events, which have the potential to damage maritime interests, means that, whatever their limitations, these prognostications are of real interest and need to be kept under close review. Because the forecasts and the reasoning behind them are published in good time it is possible for interested parties to decide whether these predictions should influence their thinking. This will have to be done either by their own analysis of the forecasts and their reasoning, or by seeking the guidance of independent experts.

These longer term forecasts involving both the atmosphere and the oceans lead naturally into the question of climate change (see Chapter 6). While this is often treated as a separate subject, in practice it is a logical extension of the issues tackled here. So in moving on to climate change it is helpful to keep in mind the challenges confronting seasonal forecasts. The huge range of factors that need to be considered and the intricate way they interact with one another are essential features of interpreting both seasonal forecasts and understanding predictions of climate change.

FOOTNOTES

[1]   UK Meteorological Office, Marine Observers Handbook, (1995).

[2]   Close to sea level this correction is 0·1 mb per metre. Given that on large ships the barometer could be up to 30 m above sea level, this is a significant correction.

[3]   Korevaar (1990).

[4]   Pedgely (1997).

[5] Rasmussen *et al* (1992).

[6] The forecasts of Gray's group can be found on the Internet at http://tropical.atmos.colostate.edu/forecasts/index.html

[7] Stockdale *et al.* (1998).

CHAPTER 6

CLIMATE CHANGE

"Nothing endures but change."
Heraclitus c540-480 BC.

The subject of climate change covers a huge range of issues relating to both the natural variability of the Earth's climate and every timescale and the potential climatic consequences of human activities.[1] Here we will concentrate on those areas which are relevant to maritime operations now and in the near future. This means focusing on the changes that have occurred in recent decades and how these fit in with our knowledge of longer term changes in the past. This leads into the question of whether the growing understanding of what may have caused these changes justifies the growing confidence in the forecasts of future climatic trends. So, the limitations of the forecasts need to be examined closely as the implications for longer term investments in the maritime industry are considerable.   So the objective is to identify the relevance of climatic variability to maritime operations and, in particular, how the industry must plan for fluctuations in the incidence of extreme events.

6.1. Temperature Records

The most comprehensive analysis of global climate change in the last century or so has been conducted on temperature records. While many of the early instrumental observations were maintained on land in Europe and North America, a reasonable estimate of global trends is available since the mid-nineteenth century. The overall assessment of all these observations is presented in the set of curves in Figure 6.1. Shipboard measurements represent a major component of this work. So here we will concentrate on shipboard observations.

Using temperature measurements at sea provides excellent insight into the challenges of measuring climate change. There are a large number of measurements of both air temperatures and surface water temperatures available from the middle of the 19th century. Air temperatures made during the daytime are now regarded as unreliable because sunlight absorbed by the decking produced exaggerated figures. But, because the difference between day and night temperatures is much smaller at sea, nightime observations provide a valuable source of climatic information.

Measurements of sea surface temperatures provide a fascinating example of how additional information can be extracted from a variety of different sources.[2] In the nineteenth century many ships kept records of the temperature of water collected in buckets over the side of the ship. To make full use of these observations corrections need to be made for whether the samples were collected in wooden or uninsulated canvas buckets, as the latter cooled more quickly. A bigger puzzle was a sudden jump in temperature of nearly 0·5 °C between 1940 and 1941. It has been established that this change was due to a sudden undocumented shift from using canvas buckets to measuring the temperature at engine intake. This change took place because during wartime using lights to make measurements over the side at night was too dangerous. As a result of a huge amount of work by various researchers on some 80 million observations it has now been concluded that observations before 1941 need to be corrected by an amount that rises from +0·13 °C in 1856, when wooden buckets were widely used, to +0·42 °C in 1940, when canvas buckets were standard. Given that the overall warming of global SSTs since the mid-nineteenth century amounts to about 0·5 °C (see Fig. 6.1.d), these corrections are a major adjustment and show how essential it is to standardise instrumental observations.

The other limitation with SST measurements is the gaps in their geographical coverage. Away from the main shipping lanes, there are many areas with few data, especially in the early part of the record. Coverage was affected by two World Wars and changing trade routes (e.g. the opening of the Suez and Panama canals in 1869 and 1914 respectively). Even in recent decades there is virtually no data for the Southern Ocean south of 45°S. While additional records may be discovered, there is no prospect of many of these gaps being filled in. So all that can be done is to extrapolate the available data on the basis of known temperature patterns to produce an agreed series for global temperatures. But the fact there are large gaps in the early part of the record must not be overlooked when interpreting the evidence of global warming.

After detailed analysis of the available data, the overall conclusion is that global near-surface temperatures have risen between 0·3 and 0·6 °C since the late nineteenth century (Fig. 6.1.c).[3] The warming observed in observations at sea is slightly less than detected over the land (Fig. 6.1.d). But the overall trend is broadly similar, and when these two sets of measurements are combined they produce the agreed curve for global warming (Fig. 6.1.c). This analysis also shows broadly the same trend for both hemispheres (Fig. 6.1.a and 6.1.b), and confirms that the warming is

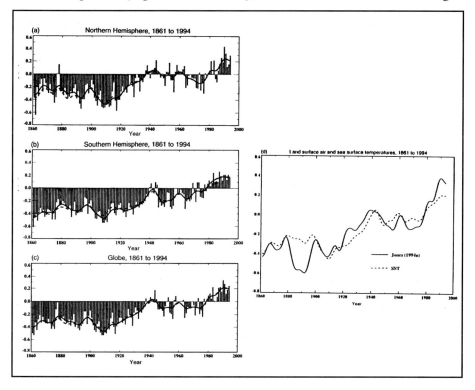

Fig 6.1
The evidence of global warming since the middle of the nineteenth century can be seen in the combined annual surface air and sea surface temperature (°C) anomalies relative to the 1961-90 average (bars and solid smoothed curves) for (a) the northern hemisphere, (b) the southern hemisphere and (c) the entire globe. The dashed curves are earlier estimates published in IPCC (1992). (d) Global land-surface air temperature (solid line) and sea surface temperature (broken line). (From IPCC, 1995, Fig. 3.3.)

a global phenomenon. The differences in part reflect the limitations of the early shipboard data, but from the beginning of the twentieth century, the estimated errors for the decadal SST anomalies are less than 0·1 °C for those areas with adequate data (about 60 per cent of the world's oceans). But when combined with the gaps in coverage this leads to the rather wide range in the quoted global warming over the last 100 years.

Equally important is the nature of recent changes. Much of the warming during the 1980s and early 1990s occurred over the northern continents, especially in winter. By comparison, the warming over the oceans was less marked, and in places, notably in the North Atlantic south of Greenland and to a lesser extent over the central North Pacific there has been some cooling. While this change is consistent with some predictions of global warming (see Section 6.4), it may be nothing more than a natural fluctuation of the climate (see Section 6.3), which could change at short notice. In particular, it appears to be closely linked to the behaviour of the North Atlantic Oscillation (NAO) (see Section 3.2). Indeed, one estimate of the impact of the NAO [4] suggests that it is responsible for about a third of the variance in the winter temperature of the northern hemisphere during the last 60 years or so.

Although temperature changes have not featured much in the discussion of maritime weather and climate, there is one industry which is dramatically effected by longer term changes in surface temperatures of the oceans. This is fishing. The best known example is the collapse of the catch of Peruvian anchovy during the El Niño of 1972. In 1970 the catch was 13 million tons (19 per cent of the world total catch of fish). This figure fell to little over 100,000 tons in the late 1970s and has not recovered. A similar collapse in cod stocks on the Grand Banks of Newfoundland in the late 1980s and early 1990s was, in part, a consequence of the sustained positive phase of the NAO. This swept exceptionally cold waters down the coast of Labrador and Newfoundland during the winter and early spring and made much of the region too cold for cod to breed. In both these cases, climatic variability has combined with overfishing and other natural fluctuations to produce a catastrophe for the fishing industry in these areas.

In other areas, knowledge of changing sea surface temperatures may be exploited by the fishing industry. For instance, the changes caused by the ENSO in the warm waters in the western equatorial Pacific affect the catches of skipjack tuna (*Katsuwonus pelamis*). Nearly 70 per cent of the world's annual tuna harvest, currently over 3 million tons, comes from the Pacific, and skipjack dominate the catch in this region. Studies of the distribution of catches of skipjack,[5] which have a sustainable catch of a million tons a year, show their movements are influenced by ENSO events and suggest the regions of highest abundance can be predicted several months in advance. So predictions of regional and global changes in sea temperature have to be included in scientific analyses of future fish stocks and what are sustainable harvesting levels.

6.2. Other Measurements

When it comes to other aspects of climate change over the oceans the problems with temperature are trivial compared with those for other parameters. Global figures for either average wind speeds and wave heights, or in the incidence of extremes are only just beginning to be extracted from available meteorological statistics, and for the most part much is built on longer term records for certain parts of the world. As has been noted in Chapter 3, those lengthy records that are available present a contradictory picture. The analysis of pressure patterns in the vicinity of the British Isles shows no significant trend in terms of either wind speeds[6] or in the incidence of gales during the winter half of the year (see Fig. 3.6).[7]

The apparently contradictory evidence on wind and waves in the North Atlantic provides a good guide to the challenges in using climatic data to plan on future changes. Even though the region has lengthier and more complete statistics than any other ocean, it is still difficult to identify clear trends. While the evidence of an upsurge in wintertime significant wave heights since the 1950s is clear cut, when viewed in the context of the variation of the NAO (see Fig. 3.7) it can be interpreted largely in terms of the fact that the NAO was at its lowest sustained level in the last 130 years in the early 1960s, and reached a very high level around 1990. The absence of wave data for the 1920s and 1930s when the NAO was in a positive phase, makes it impossible to draw conclusions about whether the upward trend of recent decades will continue. The same difficulty occurs when considering tropical cyclones. The incidence of hurricanes in the tropical Atlantic has, if anything, declined in recent decades (see Fig. 1.2. and Section 3.9). In the North Pacific there is no clear trend although there was a marked dip in the 1970s.

More generally a recently published analysis of global wind patterns over the period 1949 to 1988 suggests there has been no significant increase in the overall strength of global circulation during this period.[8] It confirms there has been an upward trend in the North Atlantic, and also in the extratropical North Pacific, which is consistent with the upsurge in winter depressions in both regions during the late 1970s and throughout the 1980s (see Section 3.2). This increase has been balanced by a downward trend in the equatorial South Atlantic, especially between July and September, and in the subtropical North Pacific. This suggests that, so far, global warming is not leading to a significant shift in atmospheric circulation strength, although regional changes may be a sign of things to come.

Weather satellites (see Section 4.4) have made a major contribution to monitoring global climate change. Two areas are of particular relevance here. One is doubt raised about the rate of global warming measured by satellite, and the other is observations of the extent of sea ice in polar regions. Even after a recent correction for orbital changes, microwave radiometer measurements of the temperature of the lower atmosphere show relative little evidence of warming since 1979 in contrast to results obtained from ground-based instruments (Fig. 6.2) although the sudden warming in 1998 has brought the two sets of observations more into line. The reason for the difference may be related to radiometers measuring upwelling radiation from a relatively thick slice of the atmosphere and so their results are not directly comparable with surface results and hence require careful calibration. Nevertheless, the satellite observations have demonstrated they are capable of making accurate measurements and the fact they produce a different warming trend since 1979 may relate to which levels of the atmosphere are being observed, and the regional nature of the warming since the 1970s. This discrepancy remains an unresolved issue in establishing the scale and causes of current global warming, and these doubts have real implications for making decisions about investments which are influenced by whether or not the climate will warm significantly in the next decade or two.

The extent of sea ice is another area where the impact of global warming has yet to show up clearly. It has fluctuated since the early 1970s (Fig. 6.3), but as yet any trends are difficult to interpret. In the Arctic it has declined by 2.9 per cent, while in the Antarctic it has expanded by about 1.3 per cent.[9] These apparently compensating changes may be in line with global warming. They are not, however, entirely consistent with the public perceptions of what is happening in the Antarctic, and provide an illuminating insight into the pitfalls of interpreting local observations, which are often used to dramatise environmental concerns. During early 1997 there was widespread concern about the break-up of the Larsen ice shelf to the east of the Antarctic Peninsula, which was presented as powerful evidence of global warming. The fact that the trend in the overall extent of Antarctic sea ice since the 1970s had been upwards was disregarded in the rush to exploit the striking photographs of the huge cracks snaking across the ice which is hundreds of metres thick (see Section 3.4). Whether the break up of the ice shelf was only a local effect resulting from shifting circulation patterns in the southern hemisphere, or whether they were the first signs of the impact of global warming - only time will tell.

This issue of identifying the cause of changes in the southern oceans was brought into sharper focus by a paper published in September 1997 by William de la Mare of the Australian Department of the Environment[10] showing an abrupt decline in the extent of Antarctic sea ice between the mid-1950s and the early 1970s. Using the logbooks that were submitted to the International Whaling Commission between 1931 and 1987, which contain about 1.5 million catch records, he was able to build up a picture of the fluctuation of the southernmost catches over the period. Because the most prized catches tended to be concentrated near the ice edge, which is an area of enhanced biological productivity, these records have proved to be an accurate measure of the extent of sea ice all around Antarctica during the summer whaling season. What they show is that the approximate mean latitude of the ice-edge remained roughly constant at about 61.5°S between 1931 and 1955 and then shifted 2.8 degrees southwards to around 64.3°S in the early 1970s, where it has remained since.

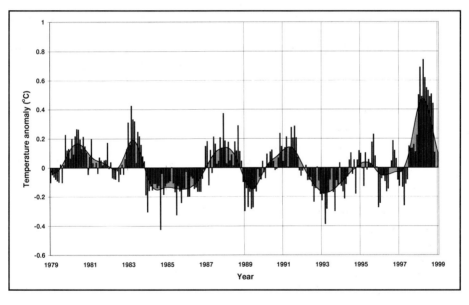

Fig 6.2
Monthly global temperature anomalies, together with a curve showing variations longer than about a year, for the lower troposphere (centered on an altitude of about 3 km [10 000 feet]) measured by microwave radiometers aboard weather satellites, showing little evidence of any warming trend in the period 1979 to 1997 but a sudden warming in 1998. (Data provided by NASA Marshall Space Flight Center, Alabama, USA.)

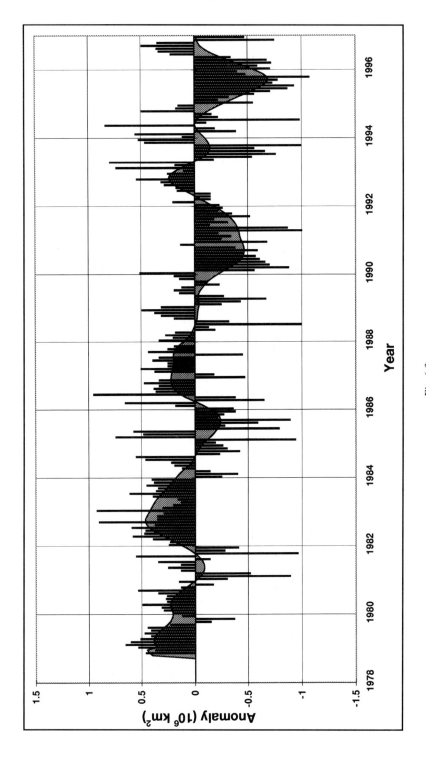

Fig 6.3a

Sea ice extent anomalies relative to 1973-1996 for (6.3a) the Northern Hemisphere and (6.3b) the Southern Hemisphere. Smooth lines show those variations of 12 months or longer. (Data from NASA Goddard Spaceflight Center, Maryland, USA).

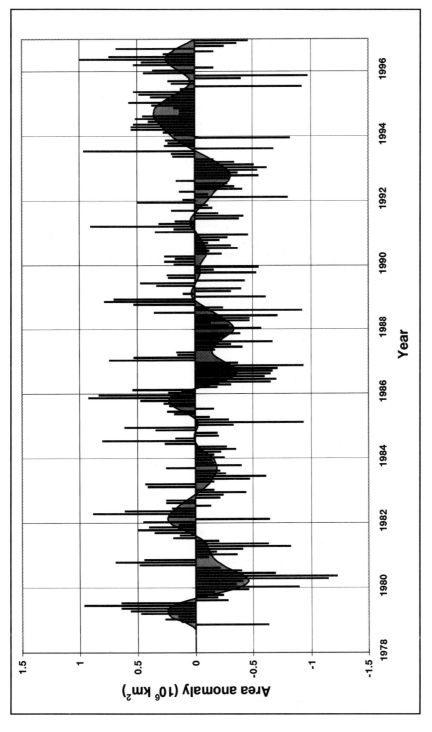

Fig 6.3b

This surprising result shows the value of proxy records of environmental change, such as ships' logbooks, in extending our knowledge of past climatic developments. It also emphasises how complicated the process of climate change is and how careful we have to be when drawing conclusions about the nature of current events. For instance, one interpretation of the break-up of the Larsen ice shelf could be that this is a regional response to the changes that took place in the extent of pack ice in the southern oceans 30 to 40 years ago. This does not mean that the response is not related to global warming, but it does underline how complicated the process of climate change is.

The challenge of getting reliable statistics applies with even greater force to identifying changes in the frequency of extreme events. So far attempts to show that global warming has produced an increase in such events have been inconclusive.[11] Although analysis of certain types of extremes across the continental United States suggests some increase in heavy precipitation events,[12] this has not been confirmed elsewhere. So, at present, the implications of the agreed observed global warming for other aspects of the climate is still the subject of debate.

What is clear is that, in spite of the lack of a discernible trend in extremes, there are significant fluctuations in their incidence on timescales from a few years to a few decades. These ups and downs are, in part, linked with changes in the oceans, notably the ENSO, which have already been touched on in Chapters 2 and 4. Now, however, we need to consider them in more detail, both to appreciate how they may have contributed to past fluctuations and also how they need to be incorporated into models of the climate if we are to have confidence in the predictions these make.

6.3. Ocean-atmosphere Interactions

The links between the oceans and the atmosphere operate in a variety of ways. Winds generate waves and, in part, drive the ocean currents (Section 2.5). Longer term fluctuations in the oceans, such as the ENSO, are both driven by the atmosphere and also modify the circulation of the atmosphere. As noted at the end of Section 4.5 the atmosphere may exert a dominant influence in the shorter term, but beyond a few years the balance may shift. So, these feedback mechanisms are the key to understanding many aspects of climate change. The shorter term aspects of the ENSO have already been covered in terms of their relevance to seasonal forecasting in Section 4.8. Here we will examine the longer term changes which may contribute to climate change. This hinges on why one or other phase of the ENSO does not become the climatic norm (see

Section 2.7), with the winds and the ocean currents flowing perpetually in one direction. Instead, as the temperature record for the central eastern equatorial Pacific (the area 150°W to 90°W and 5°N and S, often known as NINO 3) shows (Fig. 6.4) there is a pronounced quasi-cyclic fluctuation.

The explanation of the quasi-cyclic fluctuation lies in how slow-moving undulations in the thickness of the ocean surface layer slosh back and forth across the tropical Pacific. There are two types of wave. Close to the equator eastward moving changes in the depth of the thermocline (see Section 2.2), known as Kelvin waves, take two to three months to cross the Pacific. Westward moving changes in the depth of the thermocline are known as Rossby waves. These are the oceanic equivalent to the long waves observed in atmospheric circulation (see Section 2.2), and are slower moving. On the equator they take about nine months to cross the Pacific, but, they, unlike Kelvin waves are affected by the Coriolis force, because they propagate at higher latitudes, where the effects of the Earth's rotation exert a greater effect, and this means they move much more slowly at higher latitudes (at 12°N and S they take four years to cross the Pacific). When both these types of waves reach the

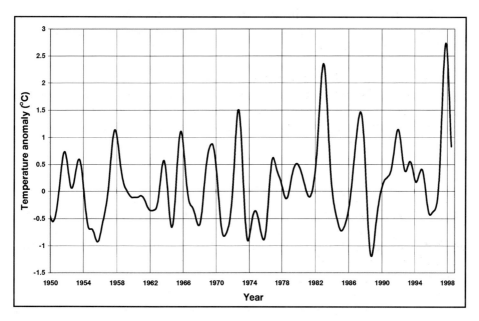

Fig 6.4
The temperature-anomaly record for the central eastern equatorial Pacific (the area 150°W to 90°W and 5°N and S, often known as NINO3) since 1950, using smoothed monthly data to highlight variations longer than about a year, showing how every few years there is a major warming with the most notable events in 1972, 1982 and 1997. (Data from NOAA, USA).

edges of the Pacific they tend to be reflected back in the direction they have come from, but they switch type, so Kelvin waves make the return as Rossby waves and vice versa. Computer models of these processes, when combined, produce realistic representations of atmospheric circulation patterns, which exhibit the property of switching back and forth between El Niño and La Niña conditions every three to five years. Moreover, if the system is subjected to a major event, such as the El Niño of 1982/83, there is a strong likelihood that it may oscillate for two or three quasi-cycles before dying away, whereas small perturbations have much less impact and lead to a more muted but chaotic response.

This ambivalent response of the models to predicting the behaviour of the ENSO provides useful insights into many aspects of the quasi-cyclic behaviour of the climate. What it tells us is that many of the apparently periodic changes in the climate may be nothing more than the chance interaction of certain of the slowly-varying components of the global weather system. This means that apparently oscillatory behaviour can build up unexpectedly but persuasively, and then, just when it begins to look like a reliable feature, suddenly fade away. This infuriating behaviour means that the moment the cycle looks good enough to make useful predictions it disappears, leading more cynical climatologists to conclude that any such cycles are untrustworthy.

The sudden appearance of the 1997 ENSO event provides further evidence that the models face an uphill task in producing realistic representations of the quasi-cyclic behaviour of the equatorial Pacific. The fact that transient atmospheric effects may have played a crucial role in triggering the event (see Section 4.8) implies that the timing of the onset may have been unpredictable. In addition there is growing climatological evidence that we cannot assume that, for the purposes of modelling work, the ocean is slowly varying, and effectively acts as a great flywheel in driving the motions of the Pacific. While there is no doubt that there is huge inertia in many of the oceans' characteristics, turbulent movements are taking place on every scale, on all timescales at all depths. So even if it transpires that the 1997 event does set the Pacific into a series of temperature ups and downs over the next decade or so, and the sudden switch to the La Niña mode during 1998 suggests that this behaviour may now be in progress, this does not mean that the system is basically predictable.

In spite of this negative conclusion it is worth recording that the three to five-year quasi-cycle associated with the ENSO does show up in many aspects of tropical meteorology. This is hardly surprising, given the dominant influence of the ENSO in the tropics (see Section 4.8), but when it occurs in its more marked oscillatory form (e.g. during 1970s and 1980s) it may be sufficiently predictable to influence decisions on timescales of several years. But, as recent events indicate, the behaviour of the ENSO has a sufficiently chaotic element to make the inclusion in planning on these timescales risky unless the stakes are low and the rewards high.

As noted in Chapter 4, the effects of the ENSO extend to higher latitudes in the North Pacific and into western North America. But it is less clear how these changes are linked with interannual fluctuations in the North Atlantic. What is certain, however, is that the NAO exhibits quasi-cyclic behaviour (see Fig. 3.7). While this is not as orderly as the ENSO, it is sufficient to exert a major influence on the climatology of wind and waves in the North Atlantic. The reason it does not match the regularity of the ENSO is that it involves not only atmosphere-ocean interactions across the Atlantic, but also effects extending over North America, Eurasia and the Arctic Ocean. The dynamics of energy transport from lower latitudes in the Atlantic together with the rapid changes in other climatically important elements such as winter snow cover and the extent of arctic pack ice make this a more chaotic combination. Nevertheless, the evidence shown in Figure 3.7 suggests that the NAO does tend to swing between times when for a number of years it is in the strong westerly mode to periods of the weaker meandering version. This behaviour does not have any dominant periodicity, being rather erratic in its behaviour, although there is some evidence of an underlying cycle of around a decade or two.

This quasi-periodic variation of the NAO may be the result of a set of feedback processes across the North Atlantic and the adjacent continents, and also with wider global atmosphere-ocean interactions. There is evidence that the position of the Gulf Stream is linked to the NAO. Measurements of the 'north wall' of the current between 1966 and 1996 show that 60 per cent of the variance of the position could be predicted in terms of the NAO.[15] Moreover, much of the remaining variance could be linked with fluctuations in the Southern Oscillation.[14] So the longer term behaviour of the North Atlantic may reflect a combination of regional feedback mechanisms and more distant influences.

This tendency for ocean-atmosphere variations to exhibit quasi-periodic behaviour of a decade or two leads to the intriguing nature of the widespread occurrence of the 20-year 'cycle' in many meteorological records.[15] It appears in global sea surface temperatures, the incidence of drought across the Great Plains of North America, the temperature records extracted from the analysis of snow deposited in the Greenland ice sheet. Although the strength of this feature in the records varies over time, the fact it appears so widely suggests that it may be a consistent feature of the climate. What is less clear is whether this faint signal is the consequence of interactions between various components of the climate system or the result of extraterrestrial influences (e.g. solar activity [sunspots] which have an 11 and 22-year cycle, or lunar tides in the atmosphere which have an 18.6-year periodicity). But, whatever its cause, it is doubtful whether its impact is sufficient to factor into the planning of maritime operations.

When it comes to longer term natural changes in the climate, the behaviour of the Great Ocean Conveyor (GOC) (see Section 2.2) is a prime candidate for causing such shifts. Because it plays a central role in the climate of the North Atlantic, if it were to have the capacity to alter its circulation, it could have a huge impact. Climatic models suggest that the amount of deep water formation and where it is formed, is extremely sensitive to the amount of freshwater entering the North Atlantic. Because this is a function of run-off from the continents, the amount of ice calved from the Greenland ice sheet, and rainfall from weather systems it can fluctuate appreciably. Models suggest that small changes in the total input may be able to trigger sudden switches to alternative conveyor belt patterns which would mean that less heat is transported to the most northern parts of the Atlantic Ocean. This could reduce the temperature of the waters around southern Greenland and Iceland by 5 °C or more. This would completely alter weather patterns around the northern hemisphere.

All this could be regarded as overly theoretical given that there is no evidence that the North Atlantic has experienced changes of this scale during the last 10,000 years. But, prior to this stable period, both during, and before and after the last Ice Age, there is clear evidence of the Atlantic circulation undergoing massive and sudden shifts in its behaviour. Against the background of the growing consensus that human activities are leading to rapid changes (see next section), this potentially unstable behaviour is an additional factor to bear in mind when interpreting the analysis of global warming. Furthermore, the fact that both changes in the NAO and the incidence of tropical storms may be attributable to shifts in surface temperatures of the Atlantic basin means the potential for switches in circulation are of interest to the maritime industry. So current research into the behaviour of global ocean circulation, both in the form of improved measurements and more sophisticated computer models, could provide valuable information on these issues. That having been said, it has to be noted that if the North Atlantic were to undergo a change of the magnitude described above, it would have overwhelming consequences.

6.4. Predicting climate change

The analysis of natural variability shows that it is not possible to make any safe predictions of how the climate may change of its own accord in the future. But, there is the additional question of how human activities could lead to global warming. This has been the subject of a huge amount of research around the world using computer models,[16] which are usually called General Circulation Models (GCMs). These models use the same set of mathematics and physics involved in the weather forecasting work described in Chapter 4. But in computing the consequences of changing certain aspects of the climate system the models have to simulate much longer periods. This has two principal consequences. First, it requires the GCMs to simulate the behaviour of the oceans and how they interact with the atmosphere. Some of these processes are included in weather forecasting, but predictions up to ten days ahead can use current sea surface temperatures and assume they effectively remain constant. Simulating the global climate for centuries ahead requires the oceans to be represented with the same realism as the atmosphere.

The second issue is simply financial. Supercomputers cost a great deal to run, and so, in the interests of keeping the computation time manageable and within budget, the models have to cut down on the detail by using a larger grid spacing and by lengthening the time intervals between successive computations. These changes are linked by the way the models operate and so reduction in resolution, both spatial and temporal, is at the cost of simplifying the representation of the climate.

Incorporating an accurate representation of the oceans into these models is particularly demanding. The major currents and the associated eddies, which are shed by them, are much smaller than corresponding atmospheric weather systems. This requires the models of the oceans to have a much higher resolution than those of the atmosphere. The highest resolution research models of the oceans have up to 60 levels and resolutions as fine as one sixth of a degree of latitude and longitude.[17] While these can produce realistic representations of many of the oceans' dynamics, they are far too expensive and slow to incorporate current climate studies. So coupled atmosphere-ocean GCMs have comparable resolution for both components.

This compromise requires careful handling. There are regions, notably in the vacinity of the major currents, where the amount of heat, momentum and water vapour exchanged between the oceans and the atmosphere varies substantially over short distances. These variations cannot be adequately resolved in existing coupled atmosphere-ocean

GCMs. As has been explained in Section 6.3, the accurate representation of these interactions is a major challenge to understanding climate change. This is particularly difficult in simulating the behaviour of the climate over many years, and requires adjustments to be made to keep the models on course. The validity of these adjustments is the subject of intense scientific debate, which will not be resolved until much more powerful models can be developed. In the meantime, these limitations in the way the oceans are represented in GCMs limit the value of their output. Furthermore, the measurements of temperature and current at depth are very sparse, even today, as compared to those for the surface layers. This means that the validation of the performance of the models of the oceans cannot be readily compared with how the oceans are actually behaving throughout their depths.

In spite of these obvious limitations, the only way forward is to try to develop better computer models to explore the behaviour of the climate. Among the most advanced examples of coupled ocean-atmosphere models is the one used by the UK Meteorological Office Hadley Centre at Bracknell, which consists of a coupled ocean-atmosphere model with a resolution of 2·5° latitude by 3·75° longitude in the atmosphere and 1·25° by 1·25° in the oceans, and the atmosphere is represented by 19 layers and the oceans by 20 layers.[18] The treatment of the dynamics of the atmosphere is effectively the same as in the weather forecasting models. It tackles the issue of the longer term adjustments of the exchanges between the atmosphere and the oceans by making calibrated seasonal flux adjustments to seasonal SSTs and ocean salinities to bring about a faithful representation of the present mean global climate. Also there is a simple treatment of sea-ice drift to remove the need for making regular flux adjustments for this important climatic parameter. In effect, the model is nudged back to reality, if it shows any tendency to drift off from what is regarded as an accurate representation of the climate, while being allowed to show the consequences of a build-up in greenhouse gases.

There are other difficulties in predicting how human activities will alter the climate. The representation of clouds is a particular challenge given the important part they play in the Earth's energy balance (see Section 2.1). Many of their natural properties still require scientific investigation, and there is also the impact of the formation of sulphate particulates resulting from the combustion of fossil fuels containing sulphur to include in the models. The predicted consequences of future human activities can vary appreciably depending on how the models handle these uncertainties. The modelling of clouds is the greatest challenge, but other features of the climate (e.g. snow cover and the extent of polar pack ice) also pose serious problems. Nonetheless, the models are making steady improvements in their ability to simulate the essential features of the global climate and continue to predict that the impact of the build-up of greenhouse gases in the atmosphere equivalent to a doubling of pre-industrial levels of CO_2 will be a rise in the global temperature of between 1.5 and 4.5 °C.[19] This build-up of greenhouse gases depends on a variety of economic and political factors, but the broad consensus is that it is likely to reach these levels in between 2050 and 2100. Moreover, any agreed international action to reduce emissions will have little impact in the next few decades, and so most of this rise is inevitable.

The Hadley Centre model shows what can be achieved with current GCMs. A recent set of results[16] gives a good idea of the scale of the computations required to form a reasonable picture and the limitations of the results obtained. It also provides a measure of the latest thinking on both the consequences of doubling the radiative forcing of CO_2 in the atmosphere, and including the effect of sulphate aerosols created by the combustion of the fossil fuels containing sulphur. The latter is particularly important as it may offer an explanation as to why the

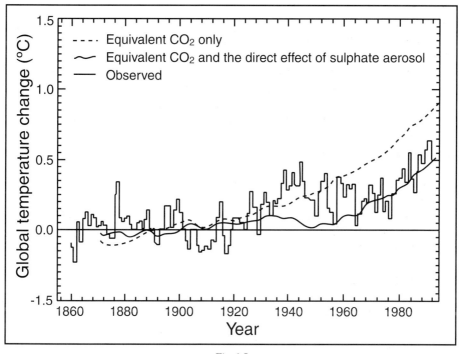

Fig 6.5
The predictions of global annual mean warming produced by the Hadley Centre, showing how the inclusion of the direct effects of sulphates (solid line) provides a better match with observations than greenhouse gases alone (dashed line). (From IPCC, 1995, Fig. 6.3.)

warming during this century, with its temporary abatement between the 1940s and the 1970s (see Section 6.1), has not followed the course predicted on the basis of the build-up of greenhouse gases alone.

The model was brought to near equilibrium through 470 years of simulated climate. After that it was run for 300 model years under control conditions during which there was no detectable trend in the global mean temperature. Thereafter three experiments were conducted each starting with the conditions in the year 1860 and running up to the year 2050. These consisted of a control with constant CO_2 concentrations, an experiment in which the concentration of CO_2 is increased gradually to give changes in radiative forcing due to all greenhouse gases, and an experiment in which both greenhouse gases and the direct radiative effect of sulphate aerosols was represented. This means that by 2050 the build-up of greenhouse gases is equivalent to doubling in CO_2 levels and the radiative impact of these dominates the climatic changes in the longer term. But the effect of sulphate aerosols is partially to counterbalance this radiative forcing especially over India, China, Mexico and Southern Africa, and to slow global warming appreciably. The results during the twentieth century (Fig. 6.5) show more realistic representation of climate change during the twentieth century. The model also predicts a somewhat less substantial warming of 2·5 °C due to doubling CO_2 levels over the period 1860 to 2050 and this warming is reduced to around 1·7 °C if the build-up of sulphate aerosols follows predicted patterns.

The impact of the predicted general warming of the global climate on maritime meteorology can only be assessed if the models give clear signals about trends in the incidence of storms and high winds at sea. Here the models are far less convincing. The simple analysis suggests two things. First, because polar regions are expected to warm more than equatorial regions, the temperature gradient and, hence, the amount of energy transported polewards will be reduced. This might be expected to reduce the strength of mid-latitude depressions in winter and also move the principal storm tracks to slightly higher latitudes. This would make shipping operations in the North Atlantic and North Pacific less hazardous in winter. The second change is the impact of rising sea surface temperatures across the tropical oceans. In principle this should lead to more frequent and intense tropical cyclones.

The objection to both these general predictions is that they are not supported by observed trends. As noted in Section 6.2 neither the trends in wind speeds, nor of tropical cyclones shows any clear evidence of the predicted rises, which might be expected, given the global warming of the last two to three decades. While this is a relatively short period in terms of detecting climate change, it does raise doubts about the ability of current climatic models to handle detailed implications of the impact of human activities even if their predictions of warming are broadly correct. Furthermore there is a worrying tendency for the different physical treatment of various climatic parameters (e.g. the properties of clouds, or sulphate particulates or snow cover) to produce contradictory regional perturbations. So there is no agreed view of how the consensus prediction of global warming will alter such basic climatic features as the strength of winter mid-latitude westerlies, the Indian monsoon or the incidence of tropical cyclones. This means that until there is a significant improvement in the models we are going to have to rely on the observation of the behaviour of the climate to establish how regional fluctuations are related to global trends.

The one area where the models do generally agree is that polar regions will warm more than lower latitudes. Moreover, these effects are likely to be greatest in winter. So one positive consequence should be a reduction of the extent of pack-ice in arctic regions although there is little evidence so far of the warming having a significant impact (see Fig. 6.3). This would increase the extent of navigable waters and reduce the hazards for any offshore structures installed in these regions.

6.5. Changes in sea level

The other aspect of global warming which is the subject of much speculation, and also of direct relevance to maritime industries, is the potential rise in sea level. Although the concept of the sea level rising globally appears simple, in practice it is not an easy thing to measure. The worldwide change in sea level is known as *eustasy*, and reflects not only thermal expansion of the oceans, but also fluctuations in the amount of water stored in groundwater, lakes and inland seas, glaciers, ice caps, and the ice sheets of Greenland and Antarctica, together with slower shifts in the Earth's crust. Because changes in the crust will differ from place to place, measurements of the eustatic rise in sea level must be sufficiently widespread to iron out the effects of local movements in the land. During the last 100 years the best estimate of the global sea level rise is 18 cm, with a range of uncertainty 10 - 25 cm.[20]

The principal factors contributing to the recent sea level rise are thermal expansion of the oceans and the melting of glaciers and ice caps around the world. Oceanic thermal expansion is a direct consequence of global warming over the last 100 years. Using coupled atmosphere-ocean computer models it is estimated that the expansion due to this warming falls in the range 2 to 6 cm with a median value of about 4 cm. The contribution from glaciers and ice caps relates to the well-documented general recession of glaciers and ice caps. Although there is considerable uncertainty about the amount of water that has been released in this melting it is estimated in total that this may have contributed about 4 cm to the sea level over the last 100 years.

Other potential contributors to the rising sea level include changes in the mass of the Antarctic and Greenland ice sheets, extraction of ground water under the continents for agricultural purposes, and changes in the volume of inland lakes and seas. In each case the analysis is equivocal, and the overall assessment is that when all factors are considered they add up to less than half the observed sea level rise. This discrepancy and the possible explanations of its cause underlines the problems of reconciling the differences and developing confidence in any projections of future rises.

Future rises in sea level depend on the detailed response of the global climate to future warming, and, in particular, precipitation patterns at high latitudes. If warming led to increased precipitation over Antarctica and Greenland then any greater peripheral melting of the ice sheets would probably be more than balanced by the accumulation of snowfall at higher levels. At lower latitudes, where glaciers and ice caps have been receding since the late nineteenth century, warming has clearly outweighed any increase in precipitation. So, for much of the coming century the rise in sea level will be dominated by the thermal expansion of the oceans plus the melting of ice caps and mountain glaciers at lower latitudes. Only, if the predicted rise in temperatures extends beyond 2100, will the melting of the ice sheets of Greenland and Antarctica become a major problem. The predicted median rise in sea level using the GCM which includes the effect of sulphate particulars is 27 cm by 2100 with a range from 17 to 49 cm.

How a rise in sea levels will affect different parts of the world also depends on how the Earth's crust will adjust to past and future changes in the load of the major ice sheets. This glacial isostatic adjustment means that the impact will vary from one coastal site to another. Superimposed on this post-glacial rebound will be regional isostatic and tectonic effects. In addition local factors such as groundwater extraction and land reclamation further complicates matters. So analysis of the specific impact of a rise in sea level will depend on a combination of improved global models of post-glacial rebound and local geological knowledge. This will involve combining what is known about local sea-level rises with climatological data to produce estimates of how the incidence of storm surges above a certain critical level will increase in the future.

An example of this type of analysis is a set of studies by Sir William Halcrow and Partners, together with the UK Meteorological Office, for the Association of British Insurers (ABI).[21]   These have analysed the vulnerability of 1300 kilometres of sea defences around the shores of England and Wales to various aspects of severe storms and the flooding they could cause. They concluded that risk assessment was an essential part of business practice and investment decisions. In terms of the insurance implications of possible upsurges in storminess, they concluded that, while insurance rates might change, but would eventually stabilise, there would be no problem in those who wished to find cover doing so, albeit at a price.

This conclusion probably applies in other countries, although the precise arrangements for insurance against flooding varies appreciably from place to place. The power of insurance premiums to influence decisions is, however, central to future developments. This potential reaction is reflected in a recent analysis of the economic costs of greenhouse-induced sea-level rises for developed property in the USA.[22] This considers a range of sea-level rises from 0.3 to 1.0 metre by the end of the next century. Depending on whether or not people exercise foresight, the costs could reach $100 million to $1 billion per annum by the year 2100.   These figures are about an order of magnitude lower than earlier estimates and reflect the inclusion of the cost-reduction potential of natural market-based adaptation in anticipation of the threat of rising seas.

The more dramatic scenarios of the melting or collapse of the Greenland and Antarctic ice sheets appear to be distant prospects. But uncertainty surrounds the stability of the West Antarctic ice sheet, which rests on a rock bed well below sea level. Depending on the physical assumptions made about this situation theoretical models can produce equivocal predictions about whether this ice sheet could collapse catastrophically if global warming and the sea level rose to some threshold value. Suffice to say, if this did happen, it could lead to a rapid rise in sea levels of several metres - far greater than anything predicted as a result of other consequences of global warming, but not something which can be included in investment decisions. This prospect has been the subject of a large number of studies[23] and the overall judgement is that possible scenarios range from the Ross Ice Shelf disintegrating over the next 200 years and the West Antarctic Ice Sheet collapsing rapidly over 50 to 200 years raising sea level 60 to 120 cm per century, to snowfall increasing and the ice sheet making a negative contribution to the sea level until global temperature has risen at least 8 °C, something that is not predicted to happen until 2200 at the earliest.

All of this suggests that in the foreseeable future sea level rises do not represent a significantly growing threat to maritime industries. In the most obvious area of the design of coastal installations, the predicted rise is not appreciably greater than that of the last 100 years. So, while it might be too complacent to interpret this as 'business as usual', it does not suggest that sea level rises are the most pressing threat to the industry - other aspects of global warming are likely to constitute a more immediate concern. Where planned installations have an expected lifetime of many decades it will, however, be necessary for investment decisions to consider how the initial design needs to accommodate the consequences of the gradual rise in sea level (e.g. the increasing risk of storm surges exceeding a certain level).

6.6. Conclusions

The lack of clarity about both the scale of any trend in global warming and also its impact on short term fluctuations in extreme weather events may be frustrating, but it does have important implications for maritime industries. First, the level of greenhouse gases in the atmosphere is bound to rise apprecibly until 2050 at least, so, if the forecasts of global warming are correct, most of the predicted warming will happen. But, the greatest changes are likely to occur initially across the northern continents. By comparison, changes over the oceans will be delayed and smaller, if not initially contradictory, given the recent cooling in the North Atlantic and North Pacific (see Section 6.1). Moreover, these changes may be complicated by the fact that they are masked by an increase in the scale of natural fluctuations from year to year and from decade to decade. This will make the interpretation of trends more difficult and place even greater emphasis on knowing what has happened in the past and how current fluctuations compare with past extremes.

In practice, this approach will require a continual appraisal of the inevitable ups and downs of the coming years. The fact that these will exhibit some degree of quasi-cyclic behaviour may be turned to good use by some far-sighted operators, but the limited ability to forecast the details of the likes of the ENSO and the NAO will make this a risky process. Nevertheless, the fact that runs of years with above or below normal levels of hurricane activity in the tropical Atlantic or intense winter depressions in the North Atlantic appear to be part of the pattern of things to come has to influence thinking.

This means that what really matters is knowing as much as possible about how the odds of extreme events may be influenced by the global climate swinging into certain patterns for a few years or more. Although the prospect of global warming may take us into uncharted waters, for the time being our knowledge must be based on the most thorough analysis of what has happened in the past. So squeezing everything possible out of old records and finding other ways of finding evidence of whether the climate can wobble more than usual for several years will be of great value. If it were possible to get a better feel of why bunches of good and bad years sometimes follow each other in close succession, it could pay handsome dividends. Until then, two basic rules seem to be worth considering. First, assume anything that has happened in the last few hundred years can occur now, although in the most extreme instances this is exceedingly unlikely. Second, if broad weather patterns start to exhibit one type of extreme, assume that it will go on behaving in this aberrant way for some time. All that can be said about the duration of any well-established anomalous seasonal patterns is they will go on until they stop, and that when they switch off it will almost certainly be unexpected.

Beyond this, if the warming trend of the 1980 and 1990's continues, and it becomes clear that human activities are contributing to a sustained global warming, past records will become increasingly less relevant. But, as far as the maritime industry is concerned, the most striking evidence of warming will occur on land. In the light of this emerging picture, it will then be a matter of deciding whether the computer models of global climate are capable of providing useful forecasts of how the incidence of various extremes will change both on land and at sea. The alternative will be simply to rely on detecting trends. Neither of these options is particularly reassuring at present. The performance of models in respect of extremes does not currently engender confidence and so it may be many years before the computer forecasts can provide what is needed. Equally, the natural variability of the climate, especially on the decadal scale, makes it hard to believe that we will get early warning of the subtle effects of global warming on the incidence of extremes. If, however, it produces a sudden and dramatic shift then this will be clear for all to see. In the meantime, we will have to draw on the past even if it starts to look out of date. But the implications of this uncertain approach are likely to be far more daunting for land-based activities, than those at sea.

FOOTNOTES

[1] The most accessible and comprehensive descriptions of the current state of research into global climate change is to be found in the various reports of the Intergovernmental Panel on Climate Change (IPCC) Reports.

[2] Folland & Parker (1995).

[3] IPCC (1995), p144.

[4] Hurrell (1996).

[5] Lehoudey *et al. (1997)*.

[6] WMO (1995), p78.

[7] Hulme & Jones (1991).

[8] Ward & Hoskins (1996).

[9] Cavalieri *et al.* (1997).

[10] de la Mare (1997).

[11] IPCC (1995), p169.

[12] Karl (1996).

[13] Taylor & Stephens (1998).

[14] Taylor *et al.* (1998).

[15] Burroughs (1994), p167.

[16] IPCC (1995), Chapter 5.

[17] IPCC (1995), p263.

[18] Mitchell *et al* (1995)

[19] IPCC (1995), p34.

[20] IPCC (1995), Chapter 7.

[21] Reeve *et al.* (1998).

[22] Yohe *et al.* (1998).

[23] Oppenheimer (1998).

CHAPTER 7

PRACTICAL APPLICATION OF ADVICE AND THE ASSESSMENT OF RISK

'Logical consequences are the scarecrows of fools and
the beacons of wise men.'
T. H. Huxley 1825-1895

So far we have explored how current knowledge of the meteorology and the climatology can be used by mariners to adapt to the behaviour of the weather and climate. In so doing we have used many examples to show how this understanding can be exploited while, at the same time, stressing the uncertainties inherent in all forecasts. Now we must draw these examples together to reach conclusions on how best to use the various forecasting services available to maritime industries to increase the safety of their operations. This involves reviewing the needs of the various sectors on timescales ranging from a few hours to several decades. In so doing, we will place particular emphasis on the interpretation of advice concerning rapidly changing conditions associated with extreme events, which pose the greatest hazards for shipping.

The emphasis on extreme events is consistent with many of the examples quoted earlier. It also reflects the simple fact that these events represent the most important threat to many maritime operations. But it is in the nature of these rare extremes that their statistical properties are hard to define. As has been stressed frequently in earlier chapters, they are not *freaks*, but represent the tail of well established statistical distributions, although the frequency of the rarest and most extreme events is difficult to quantify. This raises all sorts of problems as to whether the observations we have for any given period are representative of current conditions. In short, what meaning can be attached to, say, a 'one in a 100-year storm' and what precautions should we take to handle such an extreme, given that any such action will require additional investment.

All of these issues boil down to understanding the uncertainties in any advice and the assessment of risk. Whether we are considering the judgement of the ship's master in interpreting the signs of rapidly deteriorating weather conditions in the light of advice received from forecasters or decisions that the marine insurance industry must make about the likely incidence of catastrophes in coming decades, what matters is how much weight can be put on available meteorological knowledge, and how this then influences decisions about operations on all timescales.

7.1. Mariners

The broad conclusion is that the weather has always mattered *in all aspects of seamanship* and continues to do so. All that has changed in recent decades is that the equipment and services available to mariners have expanded dramatically. So the challenge now is to combine the standard elements of weather wisdom, which are an essential part of good seamanship, while exploiting the relevant aspects of modern meteorology to sail one's vessel more effectively and safely.

What constitutes relevant information depends on the circumstances any mariner is operating within. So it is not possible to specify what constitutes a proper level of exploitation of available forecasting services. This is probably best identified in terms of current custom and practice. This means that individual ship's masters and their mates need to form a judgement on what might normally be expected of them and then decide whether or not to match or exceed this requirement. This involves a combination of experience and formal training. What is clear, even allowing for the fundamental responsibility for the safety of the ship in the light of prevailing conditions, is that failure to demonstrate that adequate notice was taken of meteorological warnings is becoming increasingly costly and ever more difficult to justify. An essential component of this competence has to be a knowledge of basic meteorology and the strengths and weaknesses of standard forecasts. This means adequate training of staff using forecasts is a fundamental part of safe and efficient operations.

In drawing these rather obvious conclusions, we recognise that for most of the time mariners will be operating in conditions which are plain sailing and the weather plays little part in their activities. It is, however, the ability to respond effectively and in good time to deteriorating conditions, which is the mark of good seamanship. The essence of this success is the maintenance of a close eye on what the weather is doing and to be able to interpret any warning signs as and when they appear. This constant vigilance is rewarded by being better able to handle sudden bad weather.

This need for vigilance extends to time in harbour. As has been noted in a number of places the dangers of the sudden arrival of stormy weather or severe swell can endanger moored vessels. While the pressure will be on port operators to provide increasingly sophisticated specific site weather forecasts, the other side of the bargain is that ships' masters will have to make effective use of whatever information they receive. This will require the same skills as are used when at sea.

7.2. Industry

Drawing conclusions about what the maritime industry needs to know about meteorology is bound to be pretty generalised given the range of operations covered. What matters is getting the various parts of the industry to identify which aspects of meteorology impact most directly on the economics of their operations. Only by this process of self-interest can the various sectors focus on those areas which concern them most. This identification of specific interests has been an integral part of developing many specialised services. But, as has been discussed throughout this book, frequently the industry has not exploited these services to the full.

At the most basic level, we have seen that weather forecasts are not always put to good use. Ships' owners and charterers cannot disregard these services without running the risk of being called to task for failing to deliver the performance specified in contracts or incurring damage which could have been avoided if predictions had been properly heeded. Conversely, as performance improves, forecasters will be under greater pressure to explain why their output does not anticipate damaging events. These pressures will extend to other parts of the industry, which until now have not seen their role as including the provision of high quality local forecasts (e.g. some port operators). If owners and charterers are aware of what can be achieved by using readily-available forecasts, then they can specify them as a part of any contract to use specific docking installations.

Increasingly, these issues will be the subject of expensive litigation which could cost the losers dearly. Customers for these services, therefore, need to be better informed about what underlies the forecast so that they can ensure that the product is tailored to meet their specific requirements. Conversely, forecasting services have a growing interest in getting the users to understand the probabilistic nature of their output and how this should influence decisions.

7.3. Insurance risks

The attitude of the insurance industry to catastrophic weather-related losses has undergone radical revision in the last decade or so. Before 1986 no insured loss had exceed $1 billion. Hurricane *Hugo* in 1989 was five to six times higher than this figure, while for Hurricane *Andrew* in 1992 the insured loss exceeded $15 billion. While much of this rising trend in losses is a consequence of the rapid increase in size and vulnerability of shoreline communities, the implications for maritime industries are profound. Actuaries can adjust their risk calculations to reflect the changing level of development, but they have to rely on climatological statistics built up over, say, the last 30 or 100 years to estimate what are the right premiums to charge when insuring property. As has been noted earlier, the available figures are unlikely to be a good guide to the incidence or severity of future events. So the rising level of insurance losses is generating intense pressure to find ways of improving the analysis of the occurrence of extreme events.

Current evidence does not support claims of there having been a significant increase in extreme events in recent years (see Section 6.2). The rising cost of insurance claims cited above, while real, appears to be the product of other social and economic factors (e.g. more property being built in vulnerable shoreline areas, higher levels of insurance, and the chance coincidence of a few particularly damaging events). Recent analysis of hurricane damage in the US between 1925 and 1995 supports this conclusion. While the normalisation of reported losses to reflect changing economic and social circumstances is very sensitive to accuracy of the reported damage of the original event, the overall result is that only in the early 1990s did the figures return to the levels prevailing in the 1940s to 1960s. More specifically, three hurricanes, including the 1938 storm which hit New England (Section 5.4), wreaked damage comparable with Hurricane *Andrew* in 1992, and pride of place goes to the storm that hit Miami in 1926 which is estimated to have been capable of causing more then twice that of Andrew. This means that it is probably only a matter of time before the US is hit by a hurricane which does more than $50 billion's worth of damage.

Although it has yet to happen, the potential impact of any upsurge in the incidence of extreme events is, however, a pressing concern to the insurance industry. This is one reason why the emerging role of the ENSO in multi-annual climatic variability (Sections 4.7 and 5.6), and disruption of the global climate has generated so much interest. If these events can be linked to greater climatic turbulence, then there could be some justification in the belief that improved forecasts of the ENSO will lead to better predictions of the frequency of extreme events. This is of particular relevance to the incidence and intensity of tropical storms.

The recent performance of forecasts of both the development of the ENSO and its links with hurricane activity in the Atlantic suggest that it may be a very long time before we have reliable seasonal forecasts. Nevertheless, many businesses have a real interest in this work. The fact that it involves timescales which can influence financial decisions means that it has stimulated a great deal of additional work between the insurance industry and meteorologists. Typical of this effort is the Risk Prediction Initiative (RPI)[2] through which a dozen global insurance companies and an international group of scientists consider the relationships between tropical storms, climate change and the industry's exposure to catastrophic losses.

The most immediate benefit of this type of analysis is improved public information. Even though it may be many years before we can identify a significant change in the statistics of extremes, in the meantime, it can exert influence in a variety of ways. First, it provides an important check on whether current actuarial assessments of risk accurately reflect climatological knowledge. Second, if there is a need to alter the premium structures, this must reflect the best possible analysis of the statistics. Given the tight regulation of insurance companies in many countries, it is necessary that the regulators are convinced that the interpretation of the statistics is consistent across the industry. Agreement on shared and accepted information is much more likely to produce the right price structure for the industry, which increases the likelihood of the industry being able to handle the fluctuating pressures of random rare catastrophes. Similarly, in the less regulated reinsurance market, better assessment of the risk of major losses will increase the chances of individual companies being prepared to share risk. Moreover, the establishment of agreed forecast strategies on the basis of new analysis needs to be accepted by both buyers and sellers of insurance.

At present this type of analysis is concentrating on the most expensive forms of insured loss. These are concentrated onshore in the developed world. As such they reflect only a small part of the threats confronting maritime operations. Nevertheless, the methodologies being developed for the insurance companies have the potential to be applied more widely. So shipping operators and insurers need to stay abreast of work like the RPI which offers the prospect of improving economic decision-making and the safety of operations based on the effective links between meteorological science and business.

In the case of predictions of beyond a few years, these are only likely to influence investment decisions when it becomes clear that the global circulation models (GCMs) can provide useful forecasts of changes in regional climate. Although the general prediction of global warming will impact on coastal installations through the rise in sea level, it is any sustained change in storm tracks and intensities which will have the real influence on maritime operations. While planning on sea levels rising may be a sensible precaution, it is clear GCMs (see Section 6.4) are not yet able to provide the necessary guidance on changing circulation patterns. This means that for the foreseeable future the assessment of risk will have to be based on the best possible analysis of past records.

This negative conclusion also applies to the use of GCMs to assess the likelihood of there being a significant shift in the frequency of extreme events in the future. What is needed is a reliable assessment of whether certain types of extreme will become more or less frequent. Only then will it be possible to decide whether there is a need to make a substantial shift in thinking about the investment and management of maritime operations. In the meantime, the best that we can do is to rely on existing data of past events. It follows from this that any work which can provide additional information into how the climate has changed in the past will be of potential value. So various maritime industries may wish to consider whether there are any areas of research in this area worth funding. But in doing so, they will need to define closely what it is they would like to know and whether the proposed research is likely to deliver tangible benefits. This will require a level of understanding of what are the real links between climate change and their businesses which demands a degree of in-house expertise not currently available in many organisations.

7.4. The Future

The future understanding of maritime weather and climate will combine the age-old skills of being able to read current conditions from observing the sea and the sky, and exploiting technological advances. At the same time the rising tide of regulation on safety and environmental matters, plus the costs of compensation for the consequences of major accidents, means that charterers, owners and insurers will have ever greater reason to reduce the risk of weather-related losses. This combination will place an increasing burden on all those responsible for maritime operations to be adequately informed about relevant aspects of the weather and how services are developing to provide better guidance on conditions at sea. So it will make sense to stay abreast of these developments.

Progress on forecasting is likely to be a gradual business. Bigger computers and better observations will produce improvements in the quality of forecasts. These will be combined with more realistic representations of boundary layer intersections which will draw on improved physical understanding of the exchange of energy and momentum between the atmosphere and Earth's surface. These insights will be the product of large-scale international field studies which are being conducted around the world to improve our knowledge of these basic processes. There will, however, be no sudden advances, so the need to take a realistic view about predictions, which has been stressed throughout this book, will remain an essential part of the effective use of the services. Improved telecommunications and the associated visual presentation of the output of forecasting work will provide greater opportunities to explore their full meaning but will require the users to exercise more critical expertise if they are to get the best out of what is available. This is likely to require more training.

The benefits of new technology, notably satellites, are likely to be greater in developing improved climatologies. The results obtained so far show that the ability to monitor the surface of the oceans continually on a global basis is providing new insights. The prospects of building on the observations, together with the introduction of more advanced systems for measuring wave directions and wind speeds, means that better statistics of extreme conditions will be collected in the coming years. This should enable designers, insurers and owners to plan their operations using more accurate information about the conditions they are likely to experience.

The uncertainties about the nature and pace of climate change make it more difficult to draw any conclusions about the future. While the broad consensus is that global warming will continue and this will lead to rising sea levels, the more interesting issues concerning the incidence of extreme events remain shrouded in doubt. So, until clearer evidence emerges of how regional climatic patterns are altering, and what this means for maritime operations in terms of damaging weather conditions, the industry will be best advised to function on the best available current climatologies and not to place too much reliance on predictions of future changes. In so doing, it should exercise particular caution in interpreting the more extravagant claims that any short-lived extreme is a reliable indicator of longer term radical changes in the climate. These claims must be judged against the critical evaluation of available data of longer term trends. Where the avoidance of the consequences of predicted changes can only be achieved by making costly additional investments, it is probably wise to wait a bit. Only when we experience events which cannot be regarded as being part and parcel of past statistics will it be sensible to change investment strategies to accommodate whatever conditions then appear to be the emerging new order.

FOOTNOTES

[1]   Pielke & Landsea (1998). This paper provides an excellent description of the challenges that must b e addressed in making comparisons between the impact of events separated by several decades and then drawing conclusions about the likelihood of extreme losses occurring in the future.

[2]   Michaels *et al.* (1997).

BIBLIOGRAPHY

This bibliography should be read in conjunction with the reference list. It is designed to provide additional information on the content of the principal publications relevant to maritime meteorology.

Bader, M. J. Forbes, G. S. Grant, J. R. Lilley, R. B. E. and Waters, A. J. (1995). *Images in weather forecasting*. Cambridge University Press.
    This is a comprehensive review of the practical issues of using satellite images to improve weather forecasts. It contains a huge amount of information combined with many beautiful pictures, but it is not an easy read. The principal message to draw from this authoritative work is that using satellite images to interpret forecasts requires a great deal of experience and skill-not an area for the unwary.

Burroughs, W. J. (1991). *Watching the World's Weather*. Cambridge University Press.
    Although a little dated, this book provides a basic guide to the technologies used in weather satellites, what can be measured and how this information can be put to a wide variety of uses.

Burroughs, W. J. (1994). *Weather Cycles: Real or Imaginary?* Cambridge University Press.
    This presents many aspects of the global climate in the context of whether there are regular or quasi-cyclic fluctuations in the weather. As such it provides a guide to what is known about cycles, the statistical techniques which can be used to identify these fluctuations, and what they tell us about longer term variations in the climate.

Cornish, M. M. & Ives, E. E (1997). *Maritime Meteorology*. Thomas Reed Publications.

This covers much the same ground as the Meteorological Office's standard text (see below). As such it provides a clear, concise and up-to-date of the basic meteorological aspects of weather at sea.

IPCC, (1995). *Climate Change 1995: The Science of Climate Change*. J. T. Houghton, L. G. Meira Filho, B. A. Callendar, N. Harris, A. Kattenberg & K. Maskell. (eds). Cambridge University Press, UK.

This major work, together with its earlier publications, provides the most complete single presentation of current thinking on the impact of human activities on the climate. Its comprehensive and balanced approach does mean, however, that it is not the place to find easy answers to questions about the uncertainties surrounding climate change. So, if you are approaching the subject with a desire to find out more, rather than looking for easy answers, this is as good a place as any to confront the complexities of the subject.

Korevaar (1990)

A detailed analysis of available measurements in the North Sea which provides a comprehensive climatology of the region.

Kotsch (1983). *Weather for Mariners.(3rd Edition)*.

A practical presentation of many aspects of meteorology and climatology which reflects the author's lengthy experience of marine matters.

UK Meteorological Office, (1978). *Meteorology for Mariners (3rd Edition)*. HMSO.

Crammed full of sound practical advice on all aspects of the weather at sea, this book is the place to find the basic information about maritime meteorology.

UK Meteorological Office, (1995). *The Marine Observer's Handbook (11th Edition)*. HMSO.

This book provides solid practical advice on the correct way to make observations of weather conditions at sea.

US Navy, (1977) *Maritime Climatic Atlas of the World*. US Printing Office.

A distillation of the available information about the conditions at sea round the world presented principally in graphical or tabular form which enables the reader to obtain a measure of the climatology of any given area for any month of the year.

GLOSSARY

Absorption: the process by which incident radiation is taken into a body and retained without reflection or transmission, thereby increasing the internal or kinetic energy of the molecules or atoms composing the absorbing medium.

Aerosols: particles, other than water or ice, suspended in the atmosphere. They range in radius from one hundredth to one ten-millionth of a centimetre – or 10^2 to 10^{-3} micrometres (μm). Aerosols are important as nuclei for the condensation of water droplets and ice crystals, and as participants in various atmospheric chemical reactions.

Air mass: a widespread body of air that is approximately homogeneous in its horizontal extent, particularly with reference to temperature and moisture distribution; in addition, the vertical temperature and moisture distributions are approximately the same over its horizontal extent

Albedo: the proportion of the radiation falling upon a non-luminous body which it diffusely.

Black body radiation: the radiation that is emitted by a surface which absorbs all incident radiation at all wavelengths. The wavelength-dependence of this radiation is defined by the temperature of the surface.

Blocking: a phenomenon, most often association with stationary high pressure systems in the mid-latitudes of the northern hemisphere, which produces periods of abnormal weather.

Climate: the long-term statistical average of weather conditions. Global climate represents the long-term behaviour of such parameters as temperature, air pressure, precipitation, soil moisture, runoff, cloudiness, storm activity, winds, and ocean currents, integrated over the full surface of the globe. Regional climate, analogously, are the long-term averages for geographically limited domains on the Earth's surface.

Cold front: the boundary line between advancing cold air and a mass of warm air under which the cold air pushes like a wedge.

Convection: a type of heat transfer which occurs in a fluid by the vertical motion of large volumes of the heated material by differential heating (at the bottom of the atmosphere) thus creating, locally, a less dense, more buoyant fluid.

Coriolis force: the term used to explain the fact that a moving object detached from the rotating Earth appears to an observer on Earth to be deflected by a force acting at right angles to the direction of motion. Deflection of moving objects in the Northern Hemisphere is to the right of the path of motion. Deflection in the Southern Hemisphere is to the left of the path of motion.

Cryosphere: the portion of the climate system consisting of the world's ice masses, sea ice, glaciers, and snow deposits. Snow cover on land is largely seasonal and related to atmospheric circulation. Glaciers and ice sheets are tied to global water cycles and variations of sea level. The ice sheets of Greenland and the Antarctic contain 80 per cent of the existing fresh water on the globe.

Cyclone: generally a name given to a region of low pressure. In temperate latitudes cyclones are usually spoken of as depressions and the term cyclone is taken to refer to only a 'tropical cyclone'.

Depression: a part of the atmosphere where the surface pressure is lower than in surrounding parts – often called a 'low'.

Electromagnetic radiation: the emission and propagation of electromagnetic energy from a source in the form of electric and magnetic fields, which need no medium to support them and which travel through a vacuum at the velocity of light. This radiation encompasses the entire frequency range from γ-rays to radiowaves.

El Niño Southern Oscillation (ENSO): a quasi-periodic occurrence when large-scale abnormal pressure and sea-surface temperature patterns become established across the tropical Pacific every few years.

Emissivity: the ratio of the emissive power of a surface at a given temperature to that of a black body at the same temperature and the same surroundings.

Eustasy: the worldwide global changes in sea level caused by changes in the water volume of the oceans due to the formation and melting of ice sheets, variations in the amount of freshwater in lakes, inland sea and underground reserves temperature of the water, or any changes in the volume of ocean basins induced by tectonic movements.

Eustatic: a global change in sea level.

General Circulation Models (GCMs): a computational model or representation of the Earth's climate used to forecast changes in climate or weather.

Greenhouse effect: an atmospheric process in which the concentration of atmospheric trace gases (greenhouse gases) affects the amount of radiation that escapes directly into space from the lower atmosphere. Short-wave solar radiation can pass through the clear atmosphere relatively unimpeded. But long-wave terrestrial radiation, emitted by the warm surface of the Earth, is partially absorbed and then re-emitted by certain trace gases.

Greenhouse gases: the trace gases which contribute to the greenhouse effect. The main greenhouse gases are not the major constituents of the atmosphere – nitrogen and oxygen – but water vapour (the biggest contributor), carbon dioxide, methane, nitrous oxide, and (in recent years) chlorofluorocarbons. Increases in concentrations of the latter four gases have been linked to human activity.

Hadley cell: the basic vertical circulation pattern in the tropics where moist warm air rises near the equator and spreads out north and south and descends at around 20-30° N and S.

Hurricane: the name given primarily to tropical cyclones in the West Indies and Gulf of Mexico.

Insolation (from INcoming SOLar radiATION): the solar radiation received at any particular area of the Earth's surface, which varies from region to region depending on latitude and weather.

Intertropical Convergence Zone (ITCZ): a narrow low-latitude zone in which air masses originating in the northern and southern hemispheres converge and generally produce cloudy, showery weather. Over the Atlantic and Pacific it is the boundary between the north-east and south-east trade winds. The mean position is somewhat north of the equator but over the continents the range of motion is considerable.

Jet stream: strong winds in the upper troposphere whose course is related to the major weather systems in the lower atmosphere and which tend to define the movement of these systems.

Katabatic wind: a wind created when very cold air forms in upland areas and becomes sufficiently dense to drain downhill: when part of larger scale weather patterns (e.g. the Bora or Mistral in the Mediterranean, or around Antarctica) these offshore winds can be a major hazard to shipping.

Kelvin waves: gravity-inertia waves which occur in both the atmosphere and the oceans, where either the effect of the Coriolis Force is negligible (i.e. close to the equator) or where this force is balanced by the pressure gradient. The most important examples are in the equatorial stratosphere and in the thermocline of the equatorial Atlantic and Pacific close to the equator (in both cases the waves propagate eastwards relative to the Earth).

Lapse rate: a measure of the temperature profile of the atmosphere with height – the fall of temperature in a unit height.

Latent heat of vaporization: the amount of heat absorbed or emitted during the change of state from a liquid to vapour or vice versa.

Mean sea level (MSL): the average height of the sea surface, based on hourly observation of the tide height on the open coast, or in adjacent waters that have free access to the sea. In the United States, MSL is defined a the average height of the sea surface for all stages of the tide over a nineteen-year period.

Micrometre (μm): 10^{-6} metres.

Monsoon: a seasonal reversal of wind which in the summer season blows onshore, bringing with it heavy rains, and in winter blows offshore – it is of greatest meteorological importance in southern Asia. The word is believed to be derived from the Arabic word *'mausin'*, meaning a season.

Non-linearity: the lack of direct proportionality of the input and output of a physical system.

North Atlantic Oscillation (NAO): an index of the circulation in the North Atlantic which is measured in terms of the difference in pressure between the Azores and Iceland. In winter this index tends to switch between a strong westerly flow with pressure low to the north and high in the south and the reverse: the former tends to produce above normal temperatures over much of the northern hemisphere, the latter the reverse.

Polar low: small, often intense depressions that form at high latitudes in cold arctic air masses.

Quasi-biennial oscillation (QBO): the alternation of easterly and westerly winds in the equatorial stratosphere with an interval between successive corresponding maxima of 20 to 36 months. Each new regime starts above 30 km and propagates downwards at about one kilometre a month.

Radar: the use of radio waves or microwaves to measure the distance of objects by measuring the time taken for a pulse of radiation to travel from the transmitter to the object and back to an adjacent receiver.

Radar altimeter: an instrument used on satellites to measure the height of the Earth's surface, including wave heights, the profile of currents and other large scale features of the oceans.

Radiatively active trace gases: gases, present in small quantities in the atmosphere, that absorb incoming solar radiation or outgoing infrared radiation, thus affecting the vertical temperature profile of the atmosphere. These gases include water vapour, carbon dioxide, methane, nitrous oxide, chlorofluorocarbons, and ozone.

Radiometer: an instrument which makes quantitative measurements of the amount of electromagnetic radiation falling on it in a specified wavelength interval: widely used in satellite systems to measure the properties of the atmosphere oceans and the Earth's surface.

Radio-sonde: a free balloon carrying instruments which transmit measurements of temperature, pressure, and humidity to ground by radiotelegraphy as it rises through the atmosphere.

Rossby wave: in the atmosphere a wave in the general circulation in one of the principal zones of westerly winds, characterised by large wavelength (c. 6000 km), significant amplitude (c. 3000 km) and slow movement, which can be both eastward and westward relative to the Earth. In the ocean, similar waves have a wavelength of an order of a few hundred kilometres and nearly always move westward relative to the Earth.

Scatterometer: an instrument which uses a beam of microwave radiation to measure the scattering properties of the ocean surface which can then be used to estimate the direction and size of the waves, and the associated wind field.

Significant wave height: the average height of the highest third of the waves in any sea state, which equates to visual estimates by observers of the average wave height: the average height is actually just under two thirds of the estimate made by observers.

Solar radiation: the amount of radiation or energy received from the Sun at any given point.

Spatial resolution: (in the case of satellite observations) the lower limit to which the distance between features in an image can be resolved.

Stevenson shelter: a standard housing for ground-level meteorological instruments designed to ensure that reliable shade temperatures are measured.

Stratosphere: a region of the upper atmosphere, which extends from the tropopause to about 50 kilometers above the Earth's surface, and where the temperature rises slowly with altitude. The properties of the stratosphere include very little vertical mixing, strong horizontal motions, and low water vapour content compared to the troposphere. The ozone layer is located in the stratosphere.

Swell: waves generated by distant or earlier weather systems which travel large distances with little attenuation.

Thermocline: in the ocean a region of rapidly changing temperature between the warm upper layer (the epilimnion) and the colder deeper water (the hypolimnion).

Thermohaline circulation: the deep-water circulation of the oceans driven by density contrasts due to variations in salinity and temperature.

Troposphere: the lower atmosphere, from the ground to an altitude of about 8 kms at the poles, about 12 kms in mid-latitudes, and about 16 kms in the tropics. Clouds and weather systems, as experienced by people, take place in the troposphere.

Tropopause: the boundary between the troposphere and the stratosphere.

Typhoon: a name of Chinese origin (meaning 'great wind') applied to tropical cyclones which occur in the western Pacific Ocean. They are essentially the same as hurricanes in the Atlantic and cyclones in the Bay of Bengal.

Warm front: the boundary line between advancing warm air and a mass of colder air over which it rises.

Warm sector: in the early stages of the life of many depressions in temperate latitudes, there is a sector of warm air, which disappears as the system deepens and the cold front catches up the warm front.

REFERENCES

Bacon, S. and Carter, D. J. T. (1991) Wave climate changes in the North Atlantic and North Sea. *Int. J. Climatol.* **11,** 545-558.

Bader, M. J. Forbes, G. S. Grant, J. R. Lilley, R. B. E. and Waters, A. J. (1995). *Images in weather forecasting.* Cambridge University Press.

Barber, N. and Ursell, F. (1948). Study of ocean swell. *Nature.* **159,** 205.

Barnston, A. G. (1995). Our improving capability in ENSO forecasting. *Weather,* **50,** 419-30.

Barstow, S. F., and Lygre, A. (1985). Extreme Atlantic depressions during winter 1982-83: effects in Norwegian waters. *Weather,* **40,** 2-10.

Boutin, J. and Etcheto, J. (1996). Consistency of Geosat, SSM/I, and ERS-1 global surface wind speeds – Comparison with in situ data. *J. Atmos.Oceanic Technol.* **13,** 183-197.

Burroughs, W. J. (1991). *Watching the World's Weather.* Cambridge University Press.

Burroughs, W. J. (1994). *Weather Cycles: Real or Imaginary?* Cambridge University Press.

Cardonne, V. J. Jensen, R. E. Resio, D. T. Swail, V. R. and Cox, A. T. (1996) Evaluation of contemporary ocean wave models in rare extreme events: The "Halloween Storm" of October 1991 and the "Storm of the Century" of March 1993. *J. Atmos. Oceanic Technol.,* **13,** 198-230.

Cavalieri, D. Gloersen, P. Parkinson, D. L. Cosimo, J. C. and Zwally, H. J. (1997). Observed hemispheric symmetry in global sea ice changes. *Science,* **278,** 1104-1106.

Charnock, H. (1985). *Recent advances in meteorology and physical oceanography.* Royal Meteorological Society. pp 67-81.

Cornish, M. M. & Ives, E. E. (1997). *Maritime Meteorology.* Thomas Reed Publications.

Crutcher, H. L. & Quayle, R. G. (1974). *Mariners' Worldwide Climatic Guide to Tropical Storms at Sea.* US Government Printing Office.

de la Mare, W. K. (1997) Abrupt mid-twentieth-century in Antarctic sea-ice extent from whaling records. *Nature,* **389,** 57-60.

Dorr, B. (1990). Bombs over the North Atlantic. *Mariners Weather Log.* **34,** No 4, 10-15.

Draper, L. (1966). 'Freak' ocean waves. *Weather,* **21,** 2-4.

Draper, L. (1991). *Wave Climate Atlas of the British Isles, Offshore Technology Report, Department of Energy.* HMSO.

Draper, L. and Bownass, T. M. (1983) Wave devastation behind Chesil Beach. *Weather,* **38,** 346-352.

Dvorak, V. F. (1975). Tropical cyclone analysis and forecasting from satellite imagery. *Mariners Weather Log.* **19,** 199-206.

Folland, C. K. and Parker, D. E. (1995). Correction of instrumental biases in historical sea surface temperature data. *Quarterly Journal Royal Meteorological Society,* **121,** 319-367.

Hasselmanan, K. et al. (1973). Measurements of wave growth and swell decay during the Joint North Sea Wave Project (JONSWAP). *Deutche Hydrographische Zeitschrift, Ergänzsheft Reihe A.* **12,** 95pp.

Hatch, M. (1985). The meteorological interpretation the 'Sabean odours' passage in *Paradise Lost. Weather,* **40,** 115-116.

Hsu, S. A. (1993). The Gulf of Mexico – breeding ground for winter storms. *Mariners Weather Log,* **37,** No 2, 4-11.

Hulme, M. & Jones, P. D. (1991). Temperatures and windiness over the United Kingdom during the winters of 1988/89 and 1989/90 compared with previous years. *Weather,* **46,** 126-36.

Hurrell, J. W. (1996). Influences of variations in extratropical wintertime teleconnections on Northern Hemisphere temperature. *Geophys. Res. Letts.* **23,** 665-668.

IPCC, (1995). *Climate Change 1995: The Science of Climate Change.* J. T. Houghton, L. G. Meira Filho, B. A. Callendar, N. Harris, A. Kattenberg & K. Maskell. (eds). Cambridge University Press, UK.

Isemer, H-J. & Lutz, H. (1985). *The Bunker Climatic Atlas of the North Atlantic Ocean.* Springer Verlag.

Janssen, P. Hansen, B. and Bidlot, J. (1997). Verification of the ECMWF wave forecasting system against buoy and altimeter data. *Weather and Forecasting,* **12,** 763-784.

Karl, T. R., Knight, R. W., Easterling, D. R.and Quayle, R. G. (1996). Indices of climate change for the United States. *Bulletin of the American Meterological Society,* **77,** 279-92.

Korevaar , C. G. (1990) *North Sea Climatology.* Kluwer Academic Press.

Kotsch, W. J. (1983) *Weather for Mariners. (3rd Ed).* US Naval Institute.

Landsea, C. W. (1993). A climatology of intense (or major) Atlantic hurricanes. *Monthly Weather Review,* **121,** 1703-13.

Landsea, C. W., Gray, W. M., Mielke, Jr. P. W. & Berry, J. K. (1994). Seasonal forecasting of Atlantic hurricane activity. *Weather,* **49,** 273-284.

Lehoudey, P. et al (1997) El Niño Southern Oscillation and tuna in the western Pacific. *Nature,* **389,** 715.

Marine Observer (1996), **66,** 134-137

Mertins, H. O. (1968) Icing on fishing vessels due to spray. *London Marine Observer,* **38,** 128-130.

Michaels, A. Malmquist, D. Knap, A. & Close, A. (1997). Climate science and insurance risk. *Nature,* **389,** 225-228.

Mitchell, J. F. B., Johns, T. C., Gregory, J. M. & Tett, S. F. B. (1995). Climate response to increasing levels of greenhouse gases and sulphate aerosols. *Nature,* **376,** 501-4.

Mitchell, J. F. B. Davis, R. A. Ingram, W. J. and Senior, C. A. (1995). On surface temperature, greenhouse gases, and aerosols: models and observations. *J. of Climate.* **8,** 2364-2386.

Morris, R. M. (1987). Letter on record low. *Weather,* **42,** 120-122.

Murray, J. L. (1992). Ice patrol – a Titanic legacy. *Mariners Weather Log.* **36,** No 1, 21-25.

Nickerson, J. W. (1993). Freak waves. *Mariners Weather Log.* **37,** No 4, 14-19.

Oppenheimer, M. (1998). Global warming and the stability of the West Antarctic Ice Sheet. *Nature,* **393,** 325-332.

Palmer, T. N. (1993). A nonlinear dynamical perspective on climate change. *Weather,* **48,** 314-25.

Palmer, T. N. & Anderson, D. L. T. (1994). The prospects for seasonal forecasting-A review paper. *Q. J. R. Meteorol. Soc.* **120,** 755-793.

Pedgely, D. E. (1997). The Fastnet storm of 1979: A mesoscale surface jet. *Weather,* **52,** 230-242.

Philander, S. G. (1990). *El Niño, La Niña and the Southern Oscillation.* Academic Press.

Pielke, R. A. & Landsea, C. W. (1998). Normalised Hurricane Damages in the United States: 1925-1995. *Weather and Forecasting,* **13,** 621-631.

Pratt, I. (1995) The storm surge of 21 February 1993. *Weather,* **50,** 42-48.

Quayle, R. G. (1986) Weather and maritime casualties. *Mariners Weather Log.* **30,** 197-200.

Rasmussen, E. A. & Aakjaer, P. D. (1992) Two polar lows affecting Denmark. *Weather,* **47,** 326-338.

Reeve, D.E., & Burgess, K.A. (1994) A method for the assessment of coastal flood risk. *IMA J. of Maths. Applied in Business and Industry,* 197-209.

Rossiter, J. (1954) The Storm Surge in the North Sea, 31 January -1 February 1953. *Phil. Trans. Roy. Soc. A,* **246,** 371-400.

Shellard, H. C. (1974) The meteorological aspects of ice accretion on ships. *Marine Science Affairs Report No. 10. WMO – No 397.*

Stockdale, T. N., Anderson, D. L. T., Alves, J. O. S., & Balmaseda, M. A. (1998) Global seasonal rainfall forecasts using a coupled ocean-atmosphere model. *Nature,* **392,** 370-373.

Taylor, A. H. Jordan, M. B. & Stephens, J. A. (1998) Gulf Stream shifts following ENSO events. *Nature,* **393,** 638.

Taylor, A. H. & Stephens, J. A. (1998) The North Atlantic Oscillation and the latitude of the Gulf Stream. *Tellus,* **50A,** 134-142.

Torrance, J. D. (1995) Some aspects of the South African coastal low and its rogue waves. *Weather,* **50,** 163-170.

Trenberth, K. E. & Hurrell, J. W. (1994)   Decadal atmosphere-ocean variations in the Pacific. *Climate Dynamics, 9,* 303-319.

UK Admiralty, (1941). *Admiralty Weather Manual 1938.* Hydrographic Department, Admiralty.

UK Admiralty, *Arctic Pilot.*

UK Government, (1969) *Command Paper 4114, Trawler Safety (Final report of Commission on Trawler Safety. Chaired by Admiral Sir Deric Holland-Martin).* HMSO.

UK Meteorological Office, (1978).   *Meteorology for Mariners (3rd Edition).* HMSO.

UK Meteorological Office, (1995).   *The Marine Observer's Handbook (11th Edition).* HMSO.

US Navy, (1977) *Maritime Climatic Atlas of the World.* US Printing Office.

Ward, M. N. and Hoskins, B. J. (1996) Near-surface wind over the global oceans 1949-1988. *J. of Climate, 9,* 1877-1895.

WASA Group, (1998) *Bulletin of the American Met. Soc. 79,* 741-760.

WMO, (1995). *Global Climate Systems Review, June 1991-November 1993.*

Wurman, J. & Winslow, J. (1998). Intense sub-kilometre-scale boundary layer rolls observed in Hurricane Fran. *Science, 280,* 555-557.

Yohe, G., Neumann, J., Marshall, P. & Ameden, P. (1996).   The economic cost of greenhouse-induced sea-level rise for developed property in the United States. *Climate Change, 32,* 387-410.

Young, I. R. & Holland, G. J. (1996). *Atlas of the Oceans: Wind and Wave Climate.* Pergamon.

INDEX

A suffix F refers to a figure or figures, and entries in italics are names of either hurricanes or ships.